From The Nation's #1 Educational Publisher K-12

Grade 6

Enrichment Math

CONTENTS

Credits:

Cover Design & Illustration: Beachcomber Studio

McGraw-Hill Consumer Products Editorial/Production Team
Vincent F. Douglas, B.S. and M. Ed.
Tracy R. Paulus
Jennifer P. Blashkiw

McGraw-Hill
Consumer Products
A Division of The McGraw-Hill Companies

Send all inquiries to:
McGraw-Hill Consumer Products
250 Old Wilson Bridge Road
Worthington, Ohio 43085

1-57768-436-2   1 2 3 4 5 6 7 8 9 10 QPD 04 03 02 01 00 99

Name _____

WHOLE NUMBERS

WAY OUT IN SPACE

In 1772, an astronomer named Bode discovered something about the distances of planets from the Sun. The first seven planets in our solar system follow a number pattern.

1. Each number here is twice as big as the one before. Finish the pattern.

       0   3   6   12   _____ _____ _____ _____

2. Next, add 4 to each number.

     4   7   10   16

Mercury   Venus   Earth   Mars   Jupiter   Saturn   Uranus

Look at the numbers that correspond to a planet. To find the distance of the planet from the sun, multiply the planet's number by 9,300,000 miles. Use your calculator and what you know about place value to find the product. Start with Earth, number 10.

10 × 9,300,000 = 93,000,000 miles from the Sun

Find the distances for the other planets.

3. Mercury   _____

4. Venus   _____

5. Earth   9,300,000 × 10 = 93,000,000 miles

6. Mars   _____

7. Jupiter   _____

8. Saturn   _____

9. Uranus   _____

What happened to the number between Mars and Jupiter? There is no planet there, but there *is* a group of *asteroids*. These rocky objects may be a planet that exploded. Also, Neptune and Pluto, the most distant planets, do not fit the pattern.

DECIMALS

CROSS THAT BRIDGE

Choose the letter of the correct standard form for each item in the left column. The letters will spell the answer to this riddle:

What is the shortest bridge in the world?

_____	**1.** six hundred and ten ten-thousandths	**R** 0.061
_____	**2.** sixty-one thousandths	**I** 0.170
_____	**3.** one hundred and seventy thousandths	**F** 0.2
_____	**4.** fifty-three thousand and eleven hundred-thousandths	**O** 0.38
_____	**5.** three hundred eighty thousandths	**N** 0.20360
_____	**6.** two hundredths	**Y** 0.7710
_____	**7.** one hundred and seventy-three thousandths	**B** 0.0610
_____	**8.** two tenths	**E** 0.771
_____	**9.** seven thousand, seven hundred and ten ten-thousandths	**G** 0.380
_____	**10.** thirty-eight hundredths	**R** 0.2036
_____	**11.** seven hundredths	**U** 0.07
_____	**12.** two thousand and thirty-six ten-thousandths	**S** 0.5301
_____	**13.** twenty thousand, three hundred and sixty hundred-thousandths	**D** 0.53011
_____	**14.** twenty hundredths	**O** 0.173
_____	**15.** five thousand, three hundred and one ten-thousandths	**O** 0.20
_____	**16.** seven hundred and seventy-one thousandths	**E** 0.02

Name

MAKING BAR GRAPHS

AFTER-SCHOOL SPECIALTIES

Five students asked other students in their school, "What is your favorite after-school activity?" Here are the answers they got and the number of students who gave each answer.

Aaron's results: 2—chess club, 1—baseball, 1—music lessons, 1—dance lessons, 3—TV

Bill's results: 3—swimming, 1—flute lessons, 2—TV, 2—computer club, 2—reading

Sarah's results: 1—piano lessons, 2—acting club, 2—running, 2—gymnastics

Kevin's results: 2—dance lessons, 1—acting club, 1—swimming, 1—baseball, 1—reading

Jenny's results: 3—reading, 1—chess club, 2—computer club, 1—running, 1—baseball

Make a graph to show this data.

1. Decide how you want to group the answers. Make a table showing your categories.

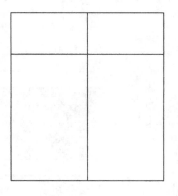

2. Decide on the labels for the two axes and make your graph.

ESTIMATING SUMS AND DIFFERENCES: ROUNDING

TARGET PRACTICE

Look at the target number in each circle. Then choose numbers
from the box whose estimated sum *or* difference will be closest to
the target. Write the addition or subtraction problem. For
problems 1–5, choose two numbers.

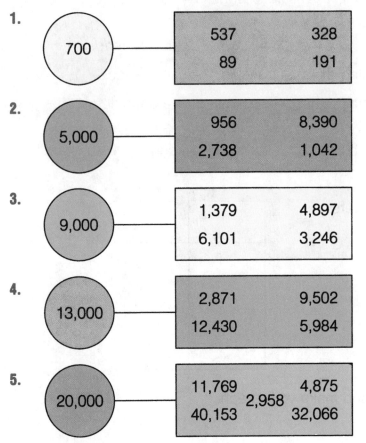

1.

700

537	328
89	191

_____

2.

5,000

956	8,390
2,738	1,042

_____

3.

9,000

1,379	4,897
6,101	3,246

_____

4.

13,000

2,871	9,502
12,430	5,984

_____

5.

20,000

11,769	4,875
--------	2,958
40,153	32,066

_____

For Problems 6–7, you may choose more than two numbers and
use more than one operation.

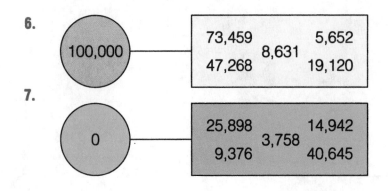

6.

100,000

73,459	5,652
--------	8,631
47,268	19,120

_____

7.

0

25,898	14,942
--------	3,758
9,376	40,645

_____

Name _____

FRONT-END ESTIMATIONS: SUMS AND DIFFERENCES

PATTERN PICKS

How would you complete the following?

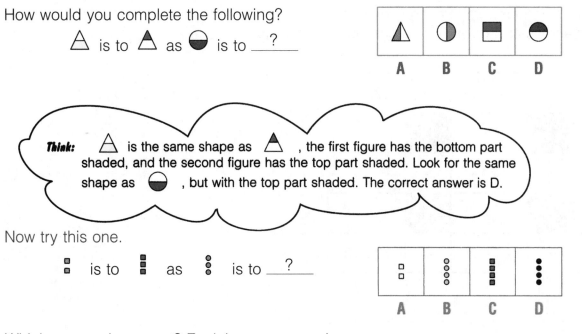

Now try this one.

Which answer is correct? Explain your reasoning. _____

_____

Write the letter of the correct answer.

		A	B	C	D
1.	△ is to ▼ as ⬜ is to ____.	▽	▶	◤	◁
2.	is to as is to ____.				
3.	L is to ⌟ as E is to ____.	Γ	⌐	⊔	⊐
4.	is to as is to ____.				
5.	is to as is to ____.				
6.	■ is to ⬛ as ● is to ____.	●	◗	◖	◠
7.	is to as is to ____.				

Adding and Subtracting Whole Numbers

SQUARE DEAL

Fill in the missing numbers in each box so the addition problems
are correct across and down. Use a calculator if you wish.

25,145	+		=	57,313
+		+		+
13,481	+	7,519	=	
=		=		=
38,626	+		=	78,313

	+	25,361	=	40,039
+		+		+
	+	10,762	=	
=		=		=
47,569	+		=	83,692

	+	26,108	=	38,306
+		+		+
	+		=	
=		=		=
23,657	+		=	81,592

Name _____

SOLVING EQUATIONS

EXTREME MEASURES

How do you measure poison ivy?

Solve each equation. Then draw a line to connect each pair of equations that have the same answer. (Use a ruler and draw a line between the dots.) The letters that are not crossed out will answer the riddle.

$n + 7 = 14$ $n =$ ___ •

$n - 3 = 6$ $n =$ ___ •

$6 + n = 12$ $n =$ ___ •

$3 + n = 8$ $n =$ ___ •

$n - 14 = 2$ $n =$ ___ •

$n - 5 = 5$ $n =$ ___ •

$n + 6 = 19$ $n =$ ___ •

$n - 7 = 4$ $n =$ ___ •

$n + 8 = 16$ $n =$ ___ •

$20 - n = 5$ $n =$ ___ •

$n + 1 = 3$ $n =$ ___ •

$n - 8 = 4$ $n =$ ___ •

$20 - n = 16$ $n =$ ___ •

$n + 5 = 19$ $n =$ ___ •

$n + 1 = 20$ $n =$ ___ •

$n - 7 = 14$ $n =$ ___ •

$n + 3 = 21$ $n =$ ___ •

o   b

y   d

a

i

t

r

f   c

n

r

h

e

s

s

e

• $6 - n = 4$ $n =$ ___

• $n + 2 = 14$ $n =$ ___

• $n - 2 = 17$ $n =$ ___

y • $n + 3 = 10$ $n =$ ___

• $n - 2 = 8$ $n =$ ___

• $n + 3 = 9$ $n =$ ___

• $n - 1 = 10$ $n =$ ___

• $n + 5 = 10$ $n =$ ___

• $n + 6 = 15$ $n =$ ___

• $n - 4 = 12$ $n =$ ___

• $n + 9 = 17$ $n =$ ___

• $16 - n = 3$ $n =$ ___

• $n + 1 = 16$ $n =$ ___

s • $5 - n = 1$ $n =$ ___

• $n - 3 = 11$ $n =$ ___

• $n - 1 = 20$ $n =$ ___

t • $n - 4 = 14$ $n =$ ___

MAKING LINE GRAPHS

COMMON QUALITIES

Here are two sets of numbers. Set A has odd numbers. Set B has even numbers:

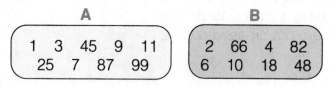

Here the two sets overlap. The numbers that fit in set C where the sets overlap have something else in common. They are under 20. One set is even and under 20; one set is odd and under 20.

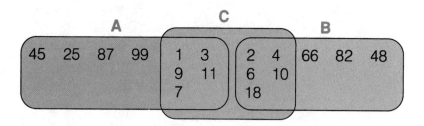

Look at each of these diagrams. Write what the numbers in sets A and B have in common.

1.

A	B
0.3 1.7 3.2 0.5 0.08 41.62 9.1	1 12 25 108 15 17 16

2.

A	B
15 91 27 1 4 16 46	101 197 125 104 118 106

For this example, write what the numbers that fit in set C have in common, too.

3.

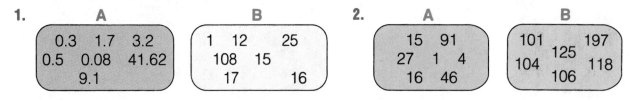

MENTAL MATH: MULTIPLYING

WHO'S WHO?

Anne, Beth, Charles, and David go to four different schools. The schools are Lincoln, Washington, Cleveland, and Roosevelt. Write the school that each student goes to. Use the clues below to help you.

Anne _____

Beth _____

Charles _____

David _____

CLUES

1. The person who goes to Washington is a boy.

2. Charles used to go to Washington, but he changed schools this year.

3. Anne's school played softball against Roosevelt last month.

4. The student at Roosevelt is a girl.

5. Charles's cousin lives on the other side of town and goes to Lincoln.

You might find it helpful to fill in this chart as you get information about the students. Write X for no and √ for yes.

	Lincoln	Washington	Cleveland	Roosevelt
Anne				
Beth				
Charles				
David				

Name

MULTIPLYING

SCRAMBLED NUMBERS

Use the numbers in the box to write the factors and the product
for each multiplication problem.

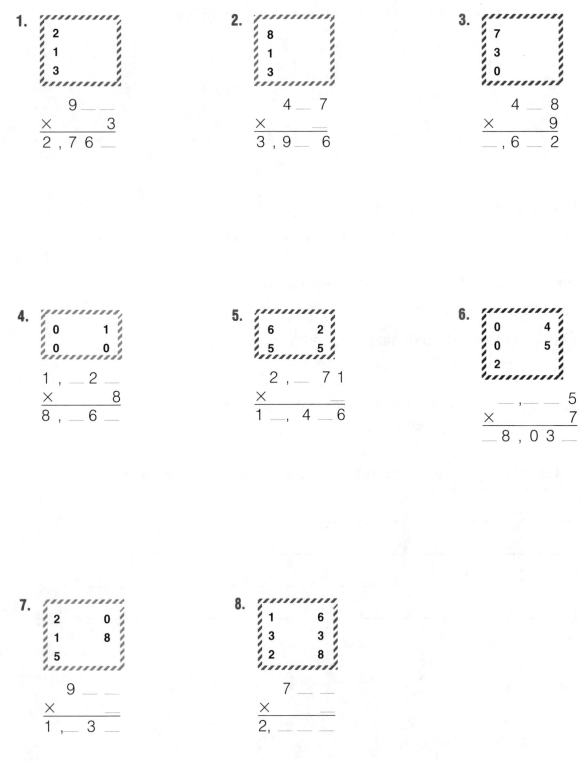

1.

| 2 |
| 1 |
| 3 |

```
        9 _ _
    ×        3
   ----------
   2 , 7 6 _
```

2.

| 8 |
| 1 |
| 3 |

```
      4 _ 7
    ×     _
   ----------
   3 , 9 _ 6
```

3.

| 7 |
| 3 |
| 0 |

```
      4 _ 8
    ×     9
   ----------
   _ , 6 _ 2
```

4.

| 0 | 1 |
| 0 | 0 |

```
   1 , _ 2 _
    ×      8
   ----------
   8 , _ 6 _
```

5.

| 6 | 2 |
| 5 | 5 |

```
     2 , _ 7 1
    ×        _
   ------------
   1 _ , 4 _ 6
```

6.

0	4
0	5
2	

```
      _ , _ _ 5
    ×         7
   ------------
   _ 8 , 0 3 _
```

7.

2	0
1	8
5	

```
        9 _ _
    ×       _
   ----------
   1 , _ 3 _
```

8.

1	6
3	3
2	8

```
       7 _ _
    ×      _
   ----------
   2 , _ _ _
```

12

MULTIPLYING GREATER NUMBERS

SERIOUS SERIES

These numbers follow a pattern. Each number is 12 times the number before it.

    3   36   432   5,184

The next two numbers in the series are 62,208 (5,184 × 12) and 746,496 (62,208 × 12).

For each series, describe the pattern. Then fill in the next two numbers in each series. Use a calculator to help you. (*Hint*: Remember, not every series uses multiplication.)

1. 4 64 1,024 _____ _____   _____

2. 5 10 15 20 25 _____ _____   _____

3. 3 51 867 _____ _____   _____

4. 23 184 1,472 11,776 _____ _____   _____

5. 3 8 40 45 225 230 _____ _____   _____

6. 6 13 91 98 686 693 _____ _____   _____

7. 12 10 660 658 43,428 _____ _____   _____

8. 19 10 90 81 729 _____ _____   _____

9. 124 248 496 992 _____ _____   _____

10. 375 75 600 300 2,400 _____ _____   _____

11. Write a series using a multiplication pattern. Give it to a friend to solve.

_____

12. Write a series with a pattern that uses addition *or* subtraction and multiplication. Give it to a friend to solve.

_____

Name _____

EXPONENTS

EXPONENTIALLY SPEAKING

Remember that you can use exponents to show multiplication when all the factors are the same.

$$5 \times 5 \times 5 \times 5 = 5^4$$

Now study the multiplication shown below.

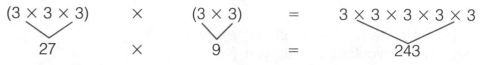

$$(3 \times 3 \times 3) \quad \times \quad (3 \times 3) \quad = \quad 3 \times 3 \times 3 \times 3 \times 3$$
$$27 \quad \times \quad 9 \quad = \quad 243$$

Write the exponents and the product in each problem. Then check that the products are correct. Use a calculator.

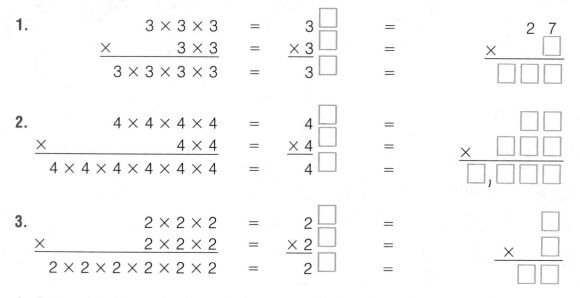

1.
$$3 \times 3 \times 3 \quad = \quad 3^{\square}$$
$$\underline{\times \qquad 3 \times 3} \quad = \quad \underline{\times 3^{\square}}$$
$$3 \times 3 \times 3 \times 3 \quad = \quad 3^{\square}$$

$$2\ 7$$
$$\underline{\times \qquad \square}$$

2.
$$4 \times 4 \times 4 \times 4 \quad = \quad 4^{\square}$$
$$\underline{\times \qquad 4 \times 4} \quad = \quad \underline{\times 4^{\square}}$$
$$4 \times 4 \times 4 \times 4 \times 4 \times 4 \quad = \quad 4^{\square}$$

3.
$$2 \times 2 \times 2 \quad = \quad 2^{\square}$$
$$\underline{\times \qquad 2 \times 2 \times 2} \quad = \quad \underline{\times 2^{\square}}$$
$$2 \times 2 \times 2 \times 2 \times 2 \times 2 \quad = \quad 2^{\square}$$

4. Can you write a rule that tells how to multiply two or more numbers in exponent form? What must be true about the base of each exponential number?

_____

_____

Use your rule to write the answer to these problems. Then write the factors and products in standard form. Check that the products are correct. Use a calculator.

5. $4^2 \times 4^2 =$ _____   _____

6. $7^4 \times 7^3 =$ _____   _____

7. $12^3 \times 12^2 =$ _____   _____

8. $2^2 \times 2^3 \times 2^4 =$ _____   _____

Name _____

MENTAL MATH: MULTIPLYING DECIMALS

MULTIPLICATION MAKES MAGIC

1. Multiply. Write each product in the box below that has the same letter as the problem. Leave **o** and **p** blank for now.

   a. 0.072 × 10 _____

   c. 0.012 × 100 _____

   e. 0.108 × 10 _____

   g. 0.0006 × 1,000 _____

   i. 0.0084 × 100 _____

   k. 0.000132 × 10,000 _____

   m. 0.0144 × 100 _____

   o. _____

   b. 0.00036 × 100 _____

   d. 0.0018 × 1,000 _____

   f. 0.0192 × 100 _____

   h. 0.048 × 10 _____

   j. 0.024 × 10 _____

   l. 0.00168 × 1,000 _____

   n. 0.000156 × 10,000 _____

   p. _____

2. In a magic square, the sum of each row, column, and diagonal should be the same. Fill in the two missing boxes so that you have a magic square. Then write multiplication problems for letters **o** and **p** above.

a.	b.	c.	d.
e.	f.	g.	h.
i.	j.	k.	l.
m.	n.	o.	p.

The magic sum is

_____ .

MULTIPLYING DECIMALS

PUZZLING NUMBERS

Multiply. Then find each answer in the grid below. Write the word
that matches each number in the grid.

1. is

$$\begin{array}{r} 4.5 \\ \times\ 0.09 \\ \hline \end{array}$$

2. funny

$$\begin{array}{r} 11.36 \\ \times\ 9.98 \\ \hline \end{array}$$

3. both

$$\begin{array}{r} 0.89 \\ \times\ 4.39 \\ \hline \end{array}$$

4. have

$$\begin{array}{r} 55.1 \\ \times\ 0.15 \\ \hline \end{array}$$

5. A

$$\begin{array}{r} 3.44 \\ \times\ 0.08 \\ \hline \end{array}$$

6. story

$$\begin{array}{r} 0.578 \\ \times\ 7.7 \\ \hline \end{array}$$

7. pencil

$$\begin{array}{r} 0.996 \\ \times\ 5.6 \\ \hline \end{array}$$

8. They

$$\begin{array}{r} 0.566 \\ \times\ 0.45 \\ \hline \end{array}$$

9. a

$$\begin{array}{r} 4.781 \\ \times\ 3.5 \\ \hline \end{array}$$

10. like

$$\begin{array}{r} 11.41 \\ \times\ 0.55 \\ \hline \end{array}$$

11. a

$$\begin{array}{r} 0.981 \\ \times\ 0.33 \\ \hline \end{array}$$

12. point

$$\begin{array}{r} 0.877 \\ \times\ 0.29 \\ \hline \end{array}$$

0.2752	5.5776	0.405	6.2755
16.7335	113.3728	4.4506	0.2547
3.9071	8.265	0.32373	0.25433

RELATING MULTIPLICATION AND DIVISION

THE BLACK BOX

The black box contains a rule that changes numbers. Look at the numbers on the left. Then see what they become when they come out of the black box.

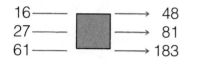

The rule inside the black box must be to multiply the number that goes in by 3.

Look at each of these examples. Write the rule. Then write the output for the last number. The rules may have more than one step, such as multiply by 3 and add 1.

1.

7 → □ → 105
1
12
14

105
15
180
___

Rule: _____

2.

2 → □ → 5
6
10
29

5
13
21
___

Rule: _____

3.

15 → □ → 38
3
10
17.5

38
14
28
___

Rule: _____

4.

1 → □ → 11
2
10
193

11
16
56
___

Rule: _____

5.

20 → □ → 6
100
28
1.2

6
26
8
___

Rule: _____

MENTAL MATH: DIVISION PATTERNS

CODE BREAKER

The letters A to J stand for the numbers 0 to 9. Two letters stand for a two-digit number. Look at the problems below. They will help you break the code. Write the number that each letter stands for.

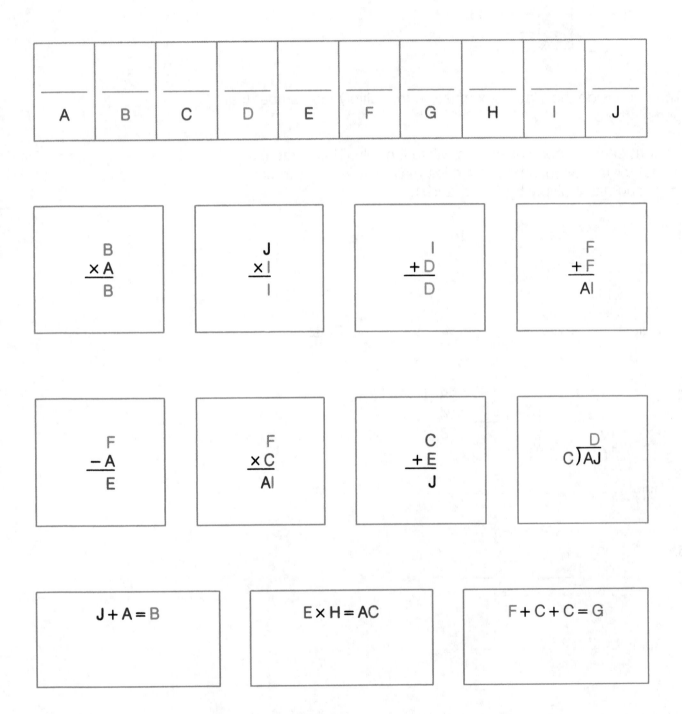

___	___	___	___	___	___	___	___	___	___
A	B	C	D	E	F	G	H	I	J

$$\begin{array}{r} B \\ \times\,A \\ \hline B \end{array} \qquad \begin{array}{r} J \\ \times\,I \\ \hline I \end{array} \qquad \begin{array}{r} I \\ +\,D \\ \hline D \end{array} \qquad \begin{array}{r} F \\ +\,F \\ \hline AI \end{array}$$

$$\begin{array}{r} F \\ -\,A \\ \hline E \end{array} \qquad \begin{array}{r} F \\ \times\,C \\ \hline AI \end{array} \qquad \begin{array}{r} C \\ +\,E \\ \hline J \end{array} \qquad C\,\overline{)\,AJ}^{\,D}$$

J + A = B   E × H = AC   F + C + C = G

DIVIDING BY ONE-DIGIT NUMBERS

PUZZLING PATTERN

Write the missing numbers and signs. Use a calculator to help you. (There is more than one right answer in some places.)

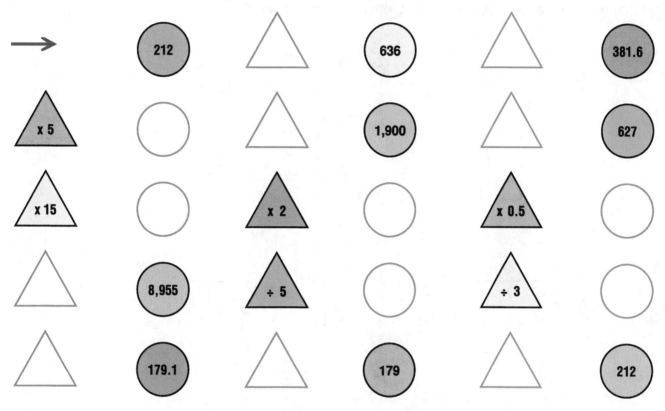

Create your own pattern puzzle in the space below.

Name _____

DIVIDING BY TWO-DIGIT NUMBERS

MISSING DIGITS

Which state is nicknamed the *Show Me State*? To find out, write
the missing digits in each division problem. Then find the
quotients below and write the correct letter in each space.

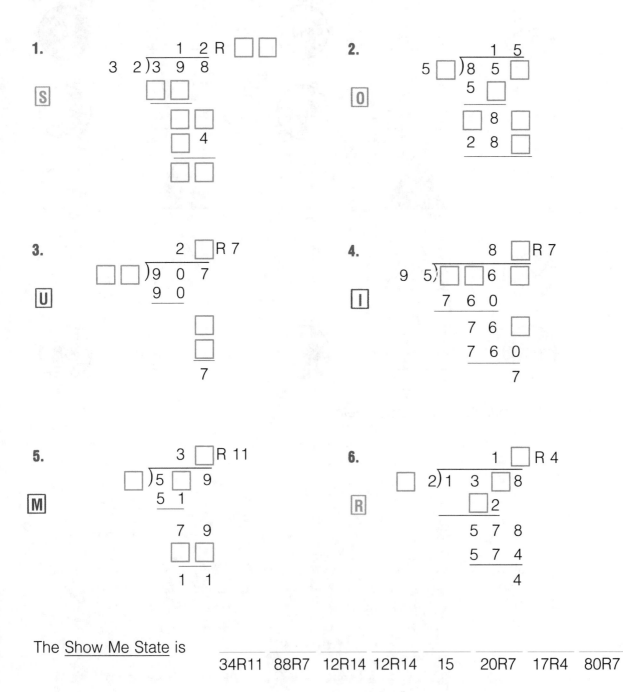

1.

S

$$
\begin{array}{r}
1\ \ 2\ R\ \square\square \\
3\ 2\,\overline{)3\ \ 9\ \ 8} \\
\square\square \\
\hline
\square\square \\
\square\ 4 \\
\hline
\square\square
\end{array}
$$

2.

O

$$
\begin{array}{r}
1\ \ 5 \\
5\,\square\,\overline{)8\ \ 5\ \ \square} \\
5\ \square \\
\hline
\square\ 8\ \square \\
2\ \ 8\ \square
\end{array}
$$

3.

U

$$
\begin{array}{r}
2\ \square\ R\ 7 \\
\square\square\,\overline{)9\ \ 0\ \ 7} \\
9\ \ 0 \\
\hline
\square \\
\square \\
7
\end{array}
$$

4.

I

$$
\begin{array}{r}
8\ \square\ R\ 7 \\
9\ 5\,\overline{)\square\square\ 6\ \square} \\
7\ \ 6\ \ 0 \\
\hline
7\ \ 6\ \square \\
7\ \ 6\ \ 0 \\
\hline
7
\end{array}
$$

5.

M

$$
\begin{array}{r}
3\ \square\ R\ 11 \\
\square\,\overline{)5\ \square\ 9} \\
5\ \ 1 \\
\hline
7\ \ 9 \\
\square\square \\
\hline
1\ \ 1
\end{array}
$$

6.

R

$$
\begin{array}{r}
1\ \square\ R\ 4 \\
\square\ 2\,\overline{)1\ \ 3\ \square\ 8} \\
\square\ 2 \\
\hline
5\ \ 7\ \ 8 \\
5\ \ 7\ \ 4 \\
\hline
4
\end{array}
$$

The <u>Show Me State</u> is ___ ___ ___ ___ ___ ___ ___ ___

34R11 88R7 12R14 12R14   15   20R7   17R4   80R7

Name

DIVIDING: CHANGING ESTIMATES

PUZZLING NUMBERS

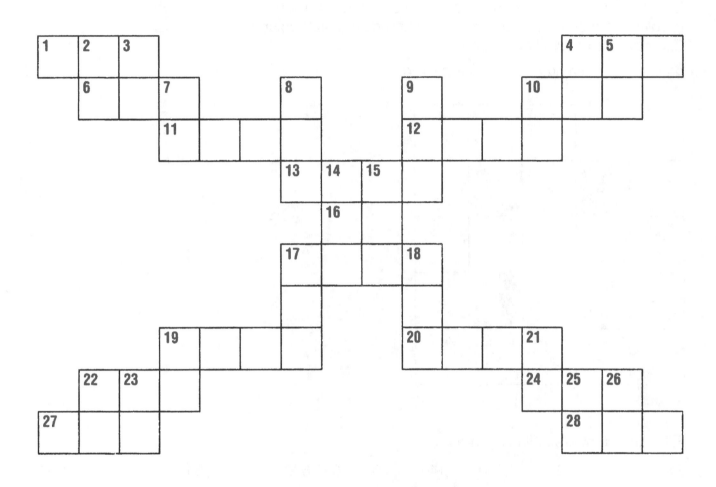

Across

1. 10,900 ÷ 25
4. ___ ÷ 7 = 79
6. ___ ÷ 4 = 157
10. ___ ÷ 2 = 52
11. ___ ÷ 60 = 26
12. ___ ÷ 47 = 58
13. ___ ÷ 88 = 52
16. ___ ÷ 7 = 8

17. ___ ÷ 4 = 711
19. ___ ÷ 25 = 200
20. ___ ÷ 82 = 28
22. 696 ÷ 6
24. ___ ÷ 4 = 131
27. ___ ÷ 7 = 94
28. ___ ÷ 10 = 23

Down

2. 540 ÷ 15
3. 5,084 ÷ 82
4. 450 ÷ 9
5. 5,292 ÷ ___ = 98
7. 2,025 ÷ 25
8. ___ ÷ 151 = 4
9. 630 ÷ 5
10. 256 ÷ 16
14. ___ ÷ 62 = 9

15. ___ ÷ 191 = 4
17. 1,600 ÷ 8
18. 1,206 ÷ 3
19. 448 ÷ 8
21. 520 ÷ 8
22. 1,380 ÷ 92
23. 36 ÷ 2
25. 990 ÷ 45
26. 2,365 ÷ 55

DIVIDING GREATER NUMBERS

SQUARE OFF

Use these numbers to fill in the square so that the division problems are correct across each row and down each column. One row is already filled in for you.

952   68   8,092   238   64,736   34

	÷		=	
÷		÷		÷
	÷		=	
=		=		=
8	÷	4	=	2

Fill in the numbers in this square.

7   4,590   595   1,080   140   20   642,600   85   54

	÷		=	
÷		÷		÷
	÷		=	
=		=		=
	÷		=	

MULTIPLICATION AND DIVISION EQUATIONS

DEEP DEPTHS

Mt. Everest is the highest point on the earth. What is the lowest point? To find out, solve each equation. Write the letter above each question in the space below that matches *n*. (Write the letter everywhere there is a matching number.)

B	**E**	**D**	**I**
$6n = 36$	$n \div 4 = 4$	$7n = 49$	$100 \div n = 10$

Y	**L**	**R**	**D**
$n \div 4 = 3$	$n \div 20 = 6$	$7n = 14$	$5n = 55$

N	**T**	**M**	**R**
$n \div 12 = 12$	$n \div 16 = 5$	$10n = 130$	$15n = 45$

A	**C**	**A**	**T**
$2n = 28$	$n \div 7 = 21$	$96 \div n = 12$	$4n = 60$

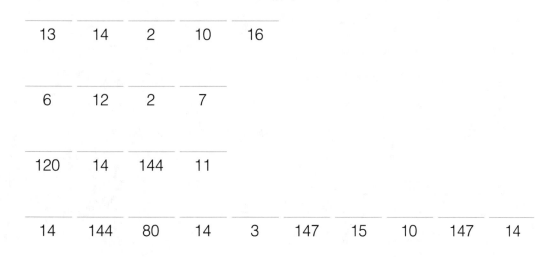

13   14   2   10   16

6   12   2   7

120   14   144   11

14   144   80   14   3   147   15   10   147   14

MENTAL MATH: DIVIDING PATTERNS WITH POWERS OF 10

A SHRINKING VIOLET?

What gets shorter as it grows older?

Divide mentally. Then arrange the answers from least to greatest on the lines below. Write the letter for each problem above it. You'll have the answer to the riddle.

L
78.2 ÷ 100 _____

R
3.2 ÷ 1,000 _____

E
14.33 ÷ 10 _____

D
67.44 ÷ 100 _____

N
0.078 ÷ 10 _____

B
0.113 ÷ 100 _____

N
122.3 ÷ 1,000 _____

I
34.6 ÷ 1,000 _____

N
566.1 ÷ 1,000 _____

C
1.8 ÷ 10 _____

U
2.03 ÷ 1,000 _____

G
14.92 ÷ 100 _____

A
45.1 ÷ 100 _____

A
0.3 ÷ 1,000 _____

_____

_____ _____ _____ _____ _____ _____ _____

_____ _____ _____ _____ _____ _____ _____

Name

ZEROS IN THE QUOTIENT AND THE DIVIDEND

ZEROED OUT

Fill in the missing numbers in these division problems.

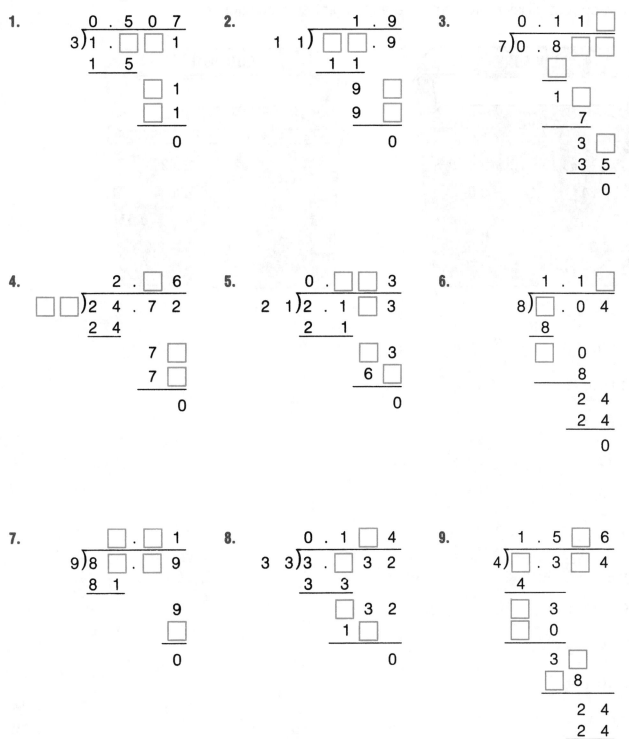

Name

DIVIDING DECIMALS BY DECIMALS

IF THE NUMBER FITS

For each problem, pick a divisor and quotient from the baskets to
make a correct division problem. Use each number exactly once.

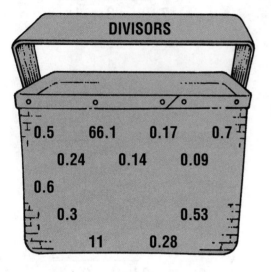

DIVISORS

0.5   66.1   0.17   0.7
    0.24   0.14   0.09
0.6
    0.3   0.53
        11   0.28

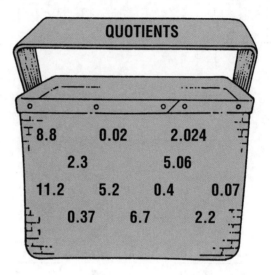

QUOTIENTS

8.8   0.02   2.024
        2.3   5.06
    11.2   5.2   0.4   0.07
        0.37   6.7   2.2

1. 0.6072 ÷ _____ = _____

2. 1.608 ÷ _____ = _____

3. 5.6   ÷ _____ = _____

4. 1.232 ÷ _____ = _____

5. 3.542 ÷ _____ = _____

6. 1.166 ÷ _____ = _____

7. 0.391 ÷ _____ = _____

8. 0.468 ÷ _____ = _____

9. 0.427 ÷ _____ = _____

10. 0.1036 ÷ _____ = _____

11. 0.012 ÷ _____ = _____

12. 4.4   ÷ _____ = _____

ROUNDING DECIMAL QUOTIENTS

UNCOVER THE FACTS

What baseball player's record for the most home runs lasted from 1935 to 1974?

To find the facts, use your ruler to connect the problems on the left with the answers on the right. Round to the nearest place indicated in parentheses. Use your calculator.

$2.5 \div 8$	(tenth)	B		R	0.398
$3.31 \div 7$	(tenth) Y	A			0.03
$0.0976 \div 8$	(hundredth)	O	C		0.144
$6.77 \div 17$	(thousandth) G		B		0.847
$0.25 \div 8$	(hundredth)	E	E		0.5
$17.8 \div 5$	(tenth)		B R		0.3
$2.55 \div 5$	(hundredth) R	U		S	0.4
$89.3 \div 6$	(thousandth) L	A	M	T	0.01
$7.62 \div 9$	(thousandth)	B V			14.883
$7.145 \div 18$	(tenth) N		E		3.6
$4.887 \div 34$	(thousandth)	H			0.51

Rearrange the letters that are not crossed out in the spaces below.

___ ___ ___ ___   ___ ___ ___ ___

Now solve this problem to find out how many home runs the player hit. Round each division step to the nearest hundredth. Then round the answer to a whole number.

$(8.273 \div 1.2) \times (459.72 \div 9) \times (16.98 \div 8.37) =$ _____

Name

ORDER OF OPERATIONS

FIGURE IT OUT

In the pattern below, there are two sets of figures. In each set,
the figures are related in the same way: The second figure is a
smaller version of the first.

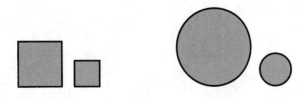

In each of the following patterns, ring the letter of the figure that completes
the second set. The figures should be related in the same way in each set.

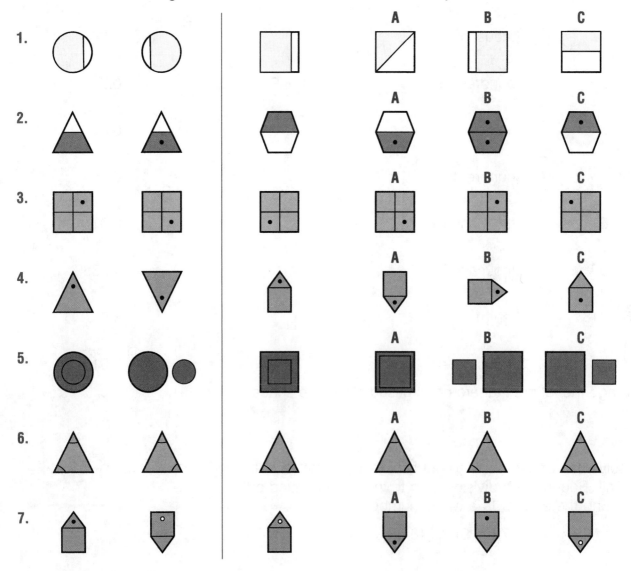

28

MEAN, MEDIAN, MODE, AND RANGE

MAKE THE GRADE

Imagine that you are teacher for a day. Here are your students' scores on a math test. There are a possible 100 points on the test. What grade will you give each student?

In order to be fair, you should use some statistics to help you. You can calculate the mean, mode, median, and range of the scores. Make up a scale telling what scores get a passing grade (A, B, C, or D) and what scores fail (F).

Student	Score	Grade
1	81	
2	95	
3	86	
4	76	
5	81	
6	94	
7	73	
8	81	
9	100	
10	52	
11	79	
12	37	
13	86	
14	75	
15	93	
16	73	
17	76	
18	90	
19	87	
20	80	

Which type of statistic was most helpful in grading the test? _____

How many of each grade did you give?

A _____ B _____ C _____ D _____ F _____

What grade did you give to a test score that was the mean on the test? _____

What grade did you give to a test score that was the median on the test? _____

What grade did you give to a test score that was the mode on the test? _____

FACTORS AND GREATEST COMMON FACTOR

DAFFY DINO DEFINITION

Which dinosaur was the fastest?

Using a ruler, draw a line from the numbers on the left to their greatest common factor (GCF) on the right. The letters that remain will answer the question. Write the letters in order, reading from top to bottom, in the blank below.

1. 12, 28 •	B P	• 15
2. 40, 4, 2 •	R	• 4
3. 75, 25, 15 •	R	• 12
4. 12, 24 •	O O N	• 20
5. 30, 40 •	N T	• 16
6. 45, 60, 75 •	T O	• 2
7. 60, 80 •	S O	• 5
8. 16, 48, 80 •	S A	• 10
9. 12, 18 •	A U	• 8
10. 14, 21 •	R R	• 6
11. 45, 72, 90 •	U S	• 9
12. 64, 24 •	U S	• 7

The fastest dinosaur was the _____

MULTIPLES AND LEAST COMMON MULTIPLE

CAREFUL ARRANGEMENTS

How can you arrange four 3s so the value is 36? Here is one way:

$$(3 + 3) \times (3 + 3) \rightarrow 6 \times 6 = 36$$

In the puzzles below, you can use any operation signs you choose, as well as decimal points or fractions.

1. Arrange four 5s so the value is 100.

2. Arrange four 9s so the value is 153.

3. Arrange four 9s so the value is 20.

4. Arrange four 5s so the value is 2.5.

5. Arrange four 3s so the value is 34.

6. Arrange five 3s so the value is 17.

7. Arrange five 2s so the value is 0.

FRACTIONS AND EQUIVALENT FRACTIONS

FRACTURED FAIRY TALE

What fairy tale is about some talking vegetables?

Write an equivalent fraction for each of the following fractions. Then find the answers under the lines below. Write the letter for each fraction on the line. The first letter has been filled in.

H $\frac{3}{8} = \frac{}{16}$

A $\frac{4}{5} = \frac{}{10}$

A $\frac{7}{8} = \frac{}{16}$

K $\frac{6}{10} = \frac{}{5}$

L $\frac{9}{10} = \frac{}{20}$

T $\frac{1}{4} = \frac{}{16}$

A $\frac{5}{10} = \frac{}{20}$

E $\frac{3}{5} = \frac{}{10}$

A $\frac{2}{4} = \frac{}{8}$

K $\frac{8}{10} = \frac{}{5}$

N $\frac{6}{8} = \frac{}{4}$

D $\frac{10}{16} = \frac{}{8}$

B $\frac{4}{10} = \frac{}{5}$

C $\frac{6}{12} = \frac{}{4}$

S $\frac{14}{20} = \frac{}{10}$

T $\frac{3}{9} = \frac{}{3}$

N $\frac{3}{6} = \frac{}{12}$

E $\frac{1}{3} = \frac{}{6}$

$\underline{\text{J}}$ $\underline{\hspace{1cm}}$ $\underline{\hspace{1cm}}$ $\underline{\hspace{1cm}}$
  $\frac{8}{10}$ $\frac{2}{4}$ $\frac{3}{5}$

$\underline{\hspace{1cm}}$ $\underline{\hspace{1cm}}$ $\underline{\hspace{1cm}}$
$\frac{14}{16}$ $\frac{3}{4}$ $\frac{5}{8}$

$\underline{\hspace{1cm}}$ $\underline{\hspace{1cm}}$ $\underline{\hspace{1cm}}$
$\frac{4}{16}$ $\frac{6}{16}$ $\frac{6}{10}$

$\underline{\hspace{1cm}}$ $\underline{\hspace{1cm}}$ $\underline{\hspace{1cm}}$ $\underline{\hspace{1cm}}$ $\underline{\hspace{1cm}}$
$\frac{2}{5}$ $\frac{2}{6}$ $\frac{4}{8}$ $\frac{6}{12}$ $\frac{7}{10}$

$\underline{\hspace{1cm}}$ $\underline{\hspace{1cm}}$ $\underline{\hspace{1cm}}$ $\underline{\hspace{1cm}}$
$\frac{1}{3}$ $\frac{10}{20}$ $\frac{18}{20}$ $\frac{4}{5}$

MIXED NUMBERS

CURIOUS CODE

In each problem, the numbers have been replaced by letters. Each letter always stands for the same number. It may be a 1-digit number or a 2-digit number. All answers are fractions or mixed numbers written in simplest terms.

Use the clue and logic to decide what number each letter stands for. Write the 12 letters and numbers on the lines below.

CLUE: L = 2

$$\frac{L}{F} = \frac{S}{L}$$
$$\frac{A}{F} = S\frac{S}{F}$$
$$\frac{E}{A} = S\frac{F}{A}$$

$$\frac{I}{E} = \frac{S}{I}$$
$$\frac{O}{F} = L\frac{I}{F}$$
$$\frac{H}{L} = I\frac{S}{L}$$

$$\frac{C}{F} = I$$
$$\frac{T}{C} = \frac{S}{L}$$
$$\frac{B}{T} = S\frac{S}{I}$$

$$\frac{Y}{F} = L\frac{S}{L}$$

Letter	Number	Letter	Number
L	2		

Now replace each number with its corresponding letter to get the answer to this riddle:

"How is a frog like a baseball player?"

___ ___ ___ ___ ___ ___ ___ ___

___ ___ ___ ___

COMPARING AND ORDERING

DAFFYNITIONS

To answer each riddle, write each group of fractions and mixed numbers in order from least to greatest. Write the letter of each number below the line.

What do you call the place where school lunches are made?

O	U	R	S	O	H	M	M
1	$\frac{2}{9}$	$\frac{2}{3}$	$\frac{1}{3}$	$\frac{7}{9}$	$\frac{4}{9}$	$\frac{1}{6}$	$1\frac{1}{3}$

___ ___ ___ ___ ___ ___ ___ ___

What do you call a twisted doughnut?

T	P	L	E	Z	E	R
$\frac{7}{12}$	$\frac{2}{12}$	$1\frac{1}{12}$	$\frac{1}{2}$	$\frac{3}{4}$	$\frac{11}{12}$	$\frac{1}{4}$

___ ___ ___ ___ ___ ___ ___

What do you call a cucumber in a sour mood?

I	L	K	P	C	E
$\frac{3}{16}$	$\frac{7}{16}$	$\frac{3}{8}$	$\frac{1}{8}$	$\frac{10}{32}$	$\frac{4}{8}$

___ ___ ___ ___ ___ ___

What do you call a song played on an automobile radio?

R	C	A	O	O	N	T
$1\frac{3}{8}$	$\frac{5}{8}$	$\frac{3}{4}$	$1\frac{3}{4}$	$1\frac{5}{8}$	$1\frac{11}{12}$	$1\frac{5}{12}$

___ ___ ___ ___ ___ ___ ___

Name

ROUNDING FRACTIONS AND MIXED NUMBERS

FRACTION ACTION

You can use mental math to decide if a fraction is greater than $\frac{1}{2}$.

Example: Is $\frac{3}{5} > \frac{1}{2}$?

Double the numerator. $2 \times 3 = 6$

If the result is greater than the denominator, $6 > 5$

the fraction is greater than $\frac{1}{2}$.   so $\frac{3}{5} > \frac{1}{2}$

Is the fraction greater than $\frac{1}{2}$? Write YES or NO.

1. $\frac{5}{9}$ _____

2. $\frac{3}{7}$ _____

3. $\frac{4}{11}$ _____

4. $\frac{7}{15}$ _____

5. $\frac{7}{10}$ _____

6. $\frac{8}{19}$ _____

7. $\frac{13}{21}$ _____

8. $\frac{11}{17}$ _____

In the maze below, find a path from START to END. Use only fractions that are greater than $\frac{1}{2}$. You may go up, down, across, or diagonally, but do not cross a number more than once.

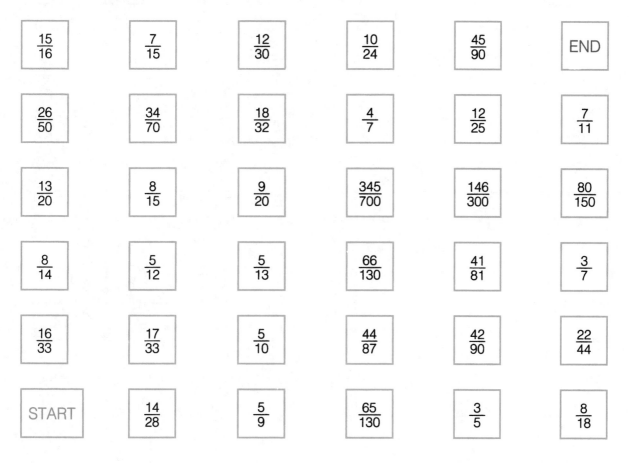

35

CUSTOMARY UNITS OF LENGTH

GOING TO GREAT LENGTH

Arrange these items from longest to shortest. (You'll have to find a way to compare their lengths first, since they are all given in different units.)

Remember, a mile is 5,280 feet.

White Sea—Baltic Canal, Soviet Union	141 mi
Golden Gate Bridge, California	1,400 yd
Simplon I Tunnel, Switzerland to Italy	12.3 mi
Sydney Harbor Bridge, Australia	19,800 in.
Panama Canal, Panama	50.7 mi
Mackinac Straits Bridge, Michigan	45,600 in.
Verrazano Narrows Bridge, New York	4,260 ft
Welland Canal, Canada	49,280 yd
Suez Canal, Egypt	100.6 mi
New River Gorge Bridge, West Virginia	1,700 ft
Rokko Tunnel, Japan	633,600 in.
Northern Line Subway Tunnel, England	30,488 yd

1. _____

2. _____

3. _____

4. _____

5. _____

6. _____

7. _____

8. _____

9. _____

10. _____

11. _____

12. _____

Name

FRACTIONS, MIXED NUMBERS, AND DECIMALS

CROSSHATCH

Try to find a decimal in columns 3 and 4 that matches each fraction or mixed number in columns 1 and 2. When you find a match, cross out both the numbers and the letters next to them. A match may be found in any row. An example is done for you.

COLUMN 1	COLUMN 2	COLUMN 3	COLUMN 4
~~C $1\frac{1}{10}$~~	A $\frac{5}{10}$	U 0.34	A 0.5
E $\frac{75}{1,000}$	P $4\frac{89}{100}$	E 6.25	T 0.11
O $\frac{1}{4}$	E $\frac{3}{4}$	B 0.25	~~D 1.1~~
T $\frac{3}{8}$	I $7\frac{99}{100}$	E 0.75	A 4.089
B $2\frac{5}{1,000}$	R $6\frac{1}{8}$	Q 9.855	E 2.005
T $\frac{6}{10}$	I $1\frac{1}{4}$	N 6.125	E 9.12
N $\frac{12}{10}$	N $9\frac{1}{2}$	A 7.099	S 7.5
E $\frac{4}{10}$	O $\frac{32}{1,000}$	M 0.4	T 1.25
H $3\frac{2}{100}$	T $\frac{7}{10}$	L 1.02	L 2.8

Now write the letters that are left in each column on the line.
Unscramble each group of letters to spell a familiar mathematical word.

COLUMN 1 _____   _____

COLUMN 2 _____   _____

COLUMN 3 _____   _____

COLUMN 4 _____   _____

ADDING FRACTIONS: UNLIKE DENOMINATORS

THE MISSING LINK

Fill in the boxes with the missing numbers. Each fraction in these problems should be in its simplest form.

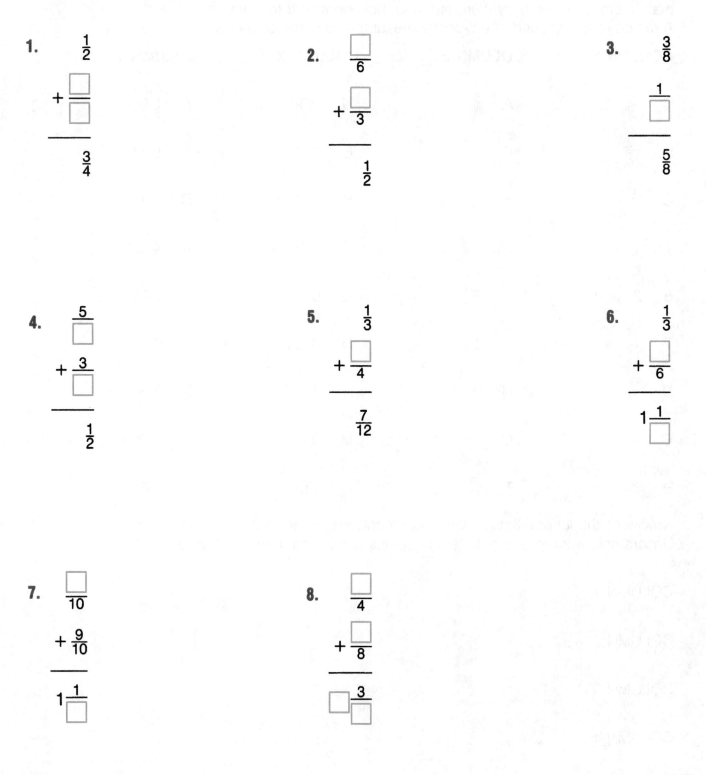

1.
$$\frac{1}{2}$$
$$+\frac{\square}{\square}$$
$$\frac{3}{4}$$

2.
$$\frac{\square}{6}$$
$$+\frac{\square}{3}$$
$$\frac{1}{2}$$

3.
$$\frac{3}{8}$$
$$\frac{1}{\square}$$
$$\frac{5}{8}$$

4.
$$\frac{5}{\square}$$
$$+\frac{3}{\square}$$
$$\frac{1}{2}$$

5.
$$\frac{1}{3}$$
$$+\frac{\square}{4}$$
$$\frac{7}{12}$$

6.
$$\frac{1}{3}$$
$$+\frac{\square}{6}$$
$$1\frac{1}{\square}$$

7.
$$\frac{\square}{10}$$
$$+\frac{9}{10}$$
$$1\frac{1}{\square}$$

8.
$$\frac{\square}{4}$$
$$+\frac{\square}{8}$$
$$\square\frac{3}{\square}$$

Name
_____

Subtracting Fractions: Unlike Denominators

PICK A PAIR

1. Which two numbers have a sum of exactly $\frac{1}{2}$?

$$\boxed{\quad \frac{1}{3} \quad \frac{1}{4} \quad \frac{1}{3} \quad \frac{1}{4} \quad \frac{1}{8} \quad \frac{1}{8} \quad}$$

_____

2. Which two numbers have a difference of $\frac{1}{8}$?

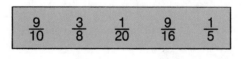

$$\frac{5}{8} \quad \frac{1}{3} \quad \frac{3}{8} \quad \frac{9}{16} \quad \frac{3}{16} \quad \frac{1}{4}$$

_____

3. Which two numbers have the sum that is closest to 1?

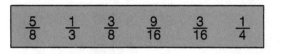

$$\frac{9}{10} \quad \frac{3}{8} \quad \frac{1}{20} \quad \frac{9}{16} \quad \frac{1}{5}$$

_____

4. Which two numbers have the difference that is closest to $\frac{1}{2}$?

$$\frac{7}{10} \quad \frac{3}{4} \quad \frac{1}{8} \quad \frac{9}{16} \quad \frac{1}{12}$$

_____

Adding Mixed Numbers

WHAT'S FOR DINNER?

What's the best kind of pie to take on a picnic?

Write the answer to each problem. Then find each answer at the bottom. Write the letter for each problem on the line to answer the riddle.

P $3\frac{1}{5}$
  $+\ 2\frac{3}{5}$

O $4\frac{1}{6}$
  $+\ 3\frac{1}{12}$

H $12\frac{1}{3}$
  $+\ 3\frac{2}{9}$

F $1\frac{3}{8}$
  $+\ 1\frac{1}{4}$

Y $4\frac{2}{9}$
  $+\ 1\frac{1}{3}$

S $9\frac{1}{5}$
  $+\ 1\frac{3}{5}$

I $1\frac{11}{12}$
  $+\ 4\frac{1}{4}$

E $1\frac{1}{6}$
  $+\ 1\frac{1}{8}$

L $3\frac{1}{4} + 2\frac{1}{4} + 6\frac{1}{2}$

O $7\frac{9}{10} + 1\frac{1}{5} + 2\frac{1}{2}$

$10\frac{4}{5}$ $15\frac{5}{9}$ $11\frac{3}{5}$ $7\frac{1}{4}$

_____   _____   _____   _____

$2\frac{5}{8}$ 12 $5\frac{5}{9}$

_____   _____   _____

$5\frac{4}{5}$ $6\frac{1}{6}$ $2\frac{7}{24}$

_____   _____   _____

SUBTRACTING MIXED NUMBERS

PART MAGIC

Fill in the missing fractions in each magic square. The sum should be the same across, down, and on the diagonals of each square.

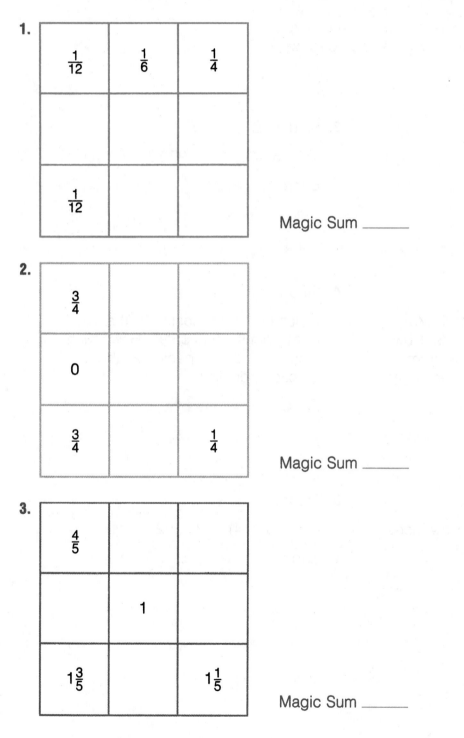

1.

$\frac{1}{12}$	$\frac{1}{6}$	$\frac{1}{4}$
$\frac{1}{12}$		

Magic Sum _____

2.

$\frac{3}{4}$		
0		
$\frac{3}{4}$		$\frac{1}{4}$

Magic Sum _____

3.

$\frac{4}{5}$		
	1	
$1\frac{3}{5}$		$1\frac{1}{5}$

Magic Sum _____

PROBLEM SOLVING

YOU BE THE TEACHER

Imagine that you are a teacher and that the problems below were solved by some of your students. Look at each problem. If the answer is correct, place a ✔ next to it. If the answer is incorrect, give the correct solution. Then explain the error so you can help your students to do better work. The first one is done for you.

1. Student A

$$
\begin{array}{r}
856 \\
723 \\
+\ 624 \\
\hline
2,193 \quad 2,203
\end{array}
$$

Error: _Did not regroup correctly._

2. Student B

145.56 + 236.3 + 99.732 = 1,166.47

Error: _____

_____

3. Student C

A box of paper plates costs $1.79 and a box of cups costs $0.79. How much change do you receive from $20 if you buy six boxes of plates and fives boxes of cups?

Error: _____ $5.31 _____

_____

4. Student D

A dozen pencils costs $2.79 and a dozen pens costs $5.95. How much more do 3 dozen pens cost than 2 dozen pencils?

Error: _____ $23.43 _____

_____

5. Student E

A box can hold 15 books. How many boxes will you need to pack 110 books?

Error: _____ 7 boxes _____

_____

6. Student F

6 × (15 + 4) − 2 = 92

Error: _____

_____

MULTIPLYING FACTORS

FLIPPER AND FRIENDS

What kind of animal is a dolphin or porpoise?

Match the multiplication problem on the left to the answer on the right. Write the letter of the answer in the box. The letters will spell the answer to the question.

1. ☐ $\frac{1}{4} \times \frac{2}{3}$ W   1

2. ☐ $\frac{3}{5} \times \frac{1}{3}$ A $\frac{7}{10}$

3. ☐ $\frac{7}{8} \times \frac{4}{5}$ A   4

4. ☐ $3 \times \frac{2}{3}$ S $\frac{1}{6}$

5. ☐ $\frac{3}{9} \times \frac{1}{4}$ H $\frac{1}{8}$

6. ☐ $4 \times \frac{1}{4}$ L $\frac{1}{12}$

7. ☐ $\frac{5}{16} \times \frac{2}{5}$ E $\frac{21}{40}$

8. ☐ $6 \times \frac{2}{3}$ L   2

9. ☐ $\frac{4}{5} \times \frac{3}{4}$ M $\frac{1}{5}$

10. ☐ $\frac{7}{10} \times \frac{3}{4}$ L $\frac{3}{5}$

MULTIPLYING MIXED NUMBERS

POSITIVELY PYRAMIDS

What numbers do cooks like?

Multiply the number at the top of each triangle by the numbers in
the circles in the second row. Write the answers in simplest form
in the bottom circles. Then arrange the answers in order from
least to greatest on the lines below. Copy the letter from each answer.

Numbers in order: ___ ___ ___ ___ ___ ___ ___ ___ ___

Letters: ___ ___ ___ ___ ___ ___ ___ ___ ___

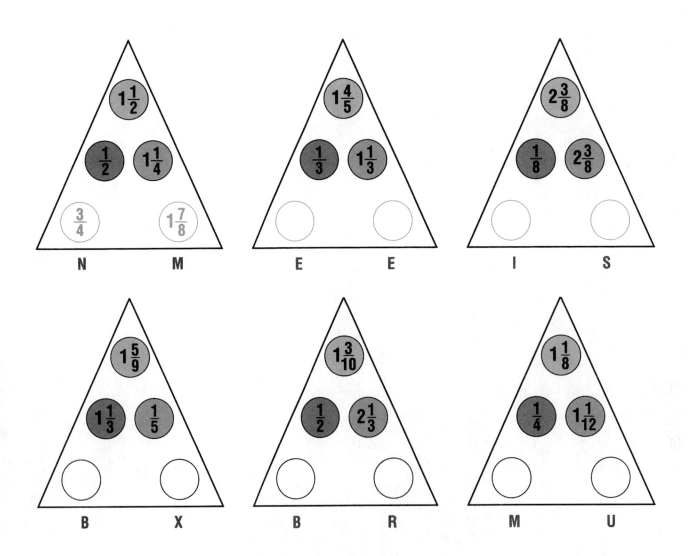

AREAS OF PARALLELOGRAMS AND TRIANGLES

AREA RUGS

At Cheap Charlie's Carpet Emporium, there are some rugs with
unusual shapes. Charlie charges $1.00 a square foot for any rug.
Figure out the total area of each rug. Write the price on the price tag.

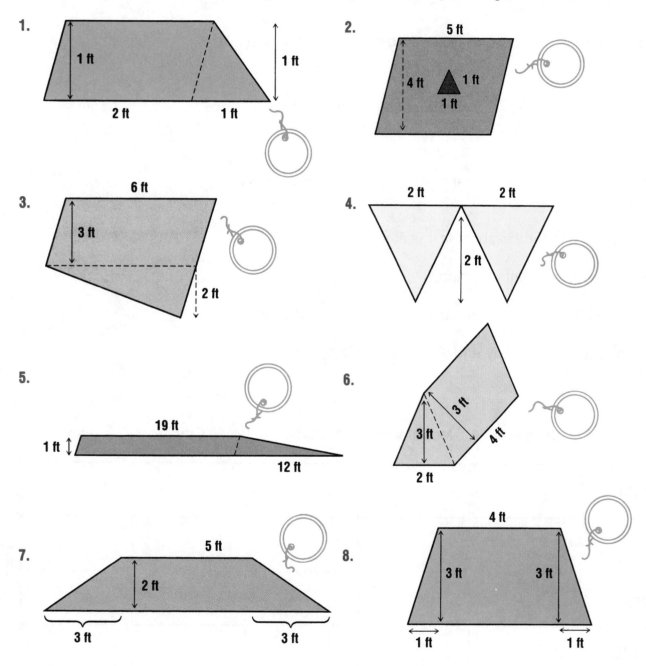

1.

1 ft   1 ft

2 ft   1 ft

2.

5 ft

4 ft   1 ft

1 ft

3.

6 ft

3 ft

2 ft

4.

2 ft   2 ft

2 ft

5.

19 ft

1 ft

12 ft

6.

3 ft

3 ft

4 ft

2 ft

7.

5 ft

2 ft

3 ft   3 ft

8.

4 ft

3 ft   3 ft

1 ft   1 ft

PROBLEM SOLVING

WHO'S WHO

Angela, Brad, Cheryl, Daniella, and Ethan each went on a vacation trip. Each went to a different place. None went to the same place he or she went to last year.

The five vacation spots were: the seashore, the mountains, a lake, a city, and a farm.

Use these clues to figure out where each person went on vacation. Fill out the chart below to help you.

- The person who went to a farm last year went to a lake this year.

- Angela went to the seashore last year.

- One of the boys went to a farm.

- Brad went to the same place that Daniella went last year.

- Ethan rode on a subway and visited tall buildings on his vacation.

	Seashore	Mountains	Lake	City	Farm
Angela					
Brad					
Cheryl					
Daniella					
Ethan					

MEASURING ANGLES

FINDING ALL THE ANGLES

Isn't this a logical idea? Suppose the numerals we use were shaped as follows: each numeral contained as many angles as the number is represented. For example, the numeral for "one" might be written this way so that it contained exactly one angle:

one angle

For each numeral below, number the angles. Show that each numeral contains the correct number of angles:

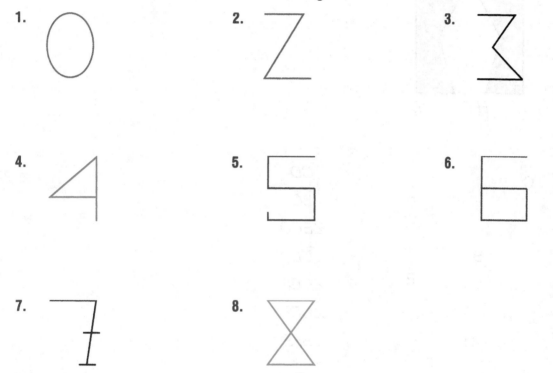

1.

2.

3.

4.

5.

6.

7.

8.

9. Can you figure out how the numeral for "nine" should be written?

PERPENDICULAR AND PARALLEL LINES

PERFECT MATCHES

What did the team call its young, impolite pitcher?

The two columns below contain a list of angles and line segments from this figure. Draw a line from each angle or line segment on the left to a congruent angle or line segment on the right. The lines will cross through some of the letters. Write the remaining letters at the bottom in the order they appear.

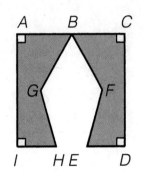

$\overline{AB}$ •

∠ABG •

∠AIH •

∠BAI •

$\overline{AI}$ •

$\overline{IH}$ •

$\overline{GH}$ •

$\overline{BG}$ •

∠BGH •

∠IHG •

• $\overline{CD}$

• $\overline{BC}$

• ∠BCD

• $\overline{CD}$

• ∠CBF

• ∠CDE

• ∠BFE

• $\overline{FE}$

• $\overline{ED}$

• ∠DEF

C B A Y G B E O L R E U S T D T E R

___ ___ ___ ___ ___ ___ ___ ___

Name

COMPASS CONSTRUCTIONS

SEEING IS BELIEVING

1. Find and circle three triangles that are the same size and shape.

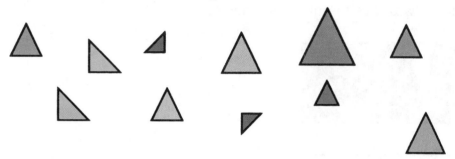

2. Find and circle two flags that have the same size poles and are pointing the same way.

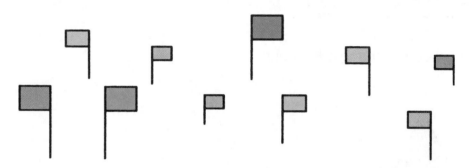

3. Find and circle two that are the same and in the same position.

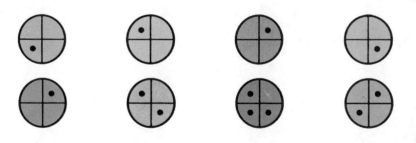

4. Find and circle two pairs that are the same shape but not the same size.

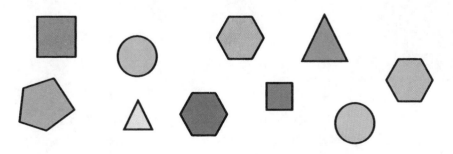

Name

QUADRILATERALS

CUTUPS

1. There are several ways to cut a rectangle into four congruent rectangles. Draw lines to show two ways to cut the rectangles.

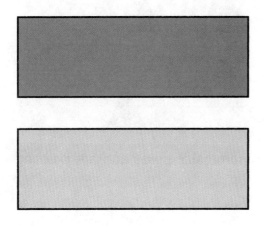

2. Draw lines to show how to divide this triangle into four congruent triangles.

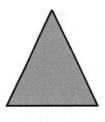

3. Draw lines to show how to divide this triangle into nine congruent triangles.

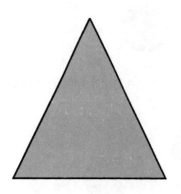

OTHER POLYGONS

COMBOS

There are three shapes on the left. Which of the figures on the right can be formed from the three shapes? The shapes can be moved or turned in any direction.

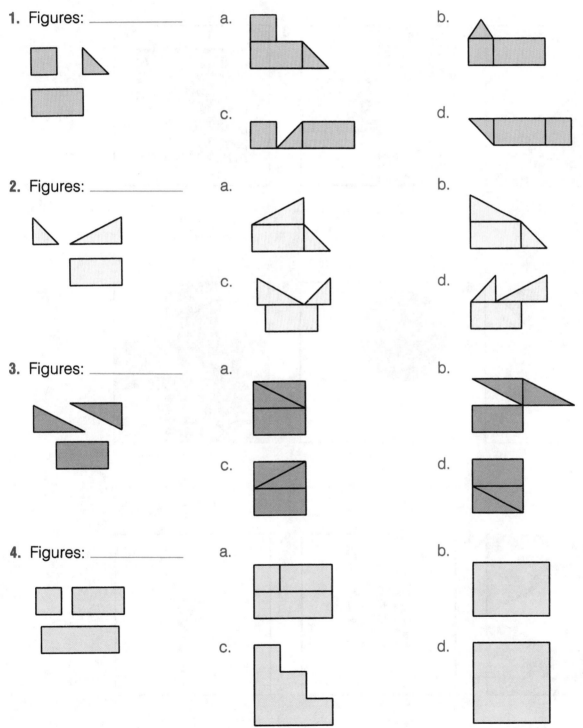

1. Figures: _____

a.

b.

c.

d.

2. Figures: _____

a.

b.

c.

d.

3. Figures: _____

a.

b.

c.

d.

4. Figures: _____

a.

b.

c.

d.

RATIOS AND RATES

ME AND MY SHADOW

Here are some patterns where shadows are seen on a sidewalk square. Estimate the ratio of light area to shadow in each square. Write the ratio.

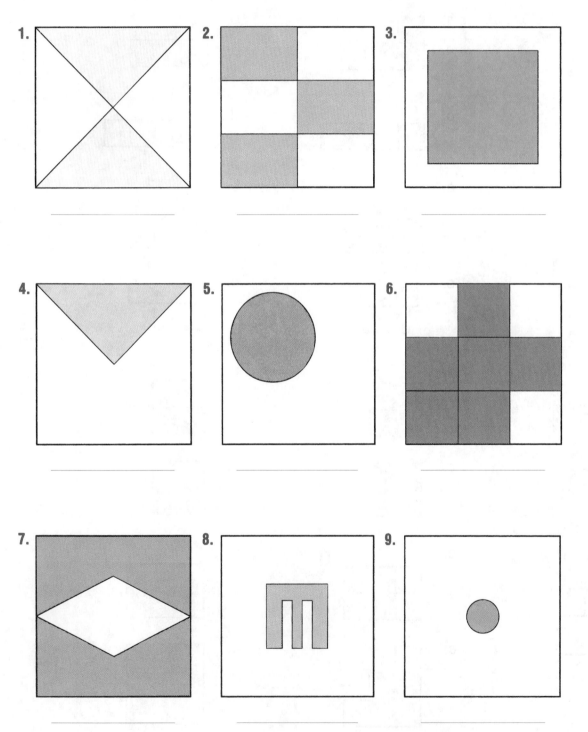

1.

2.

3.

4.

5.

6.

7.

8.

9.

PROPORTIONS

PUZZLING PATTERN

About how many volcanoes are there in the world?

To find out, solve each proportion. Write the answer. Then color in the space at the bottom with that number in it. The pattern you make will tell you the answer. Don't let your eyes trick you before you solve the problems.

1. $\dfrac{1}{8} = \dfrac{n}{16}$

2. $\dfrac{3}{5} = \dfrac{6}{n}$

3. $\dfrac{12}{n} = \dfrac{1}{3}$

4. $\dfrac{n}{27} = \dfrac{6}{18}$

5. $\dfrac{10}{100} = \dfrac{5}{n}$

6. $\dfrac{48}{n} = \dfrac{2}{4}$

7. $\dfrac{2.3}{4.6} = \dfrac{n}{6.2}$

8. $\dfrac{7}{49} = \dfrac{12}{n}$

9. $\dfrac{30}{12} = \dfrac{n}{40}$

10. $\dfrac{66}{n} = \dfrac{22}{2}$

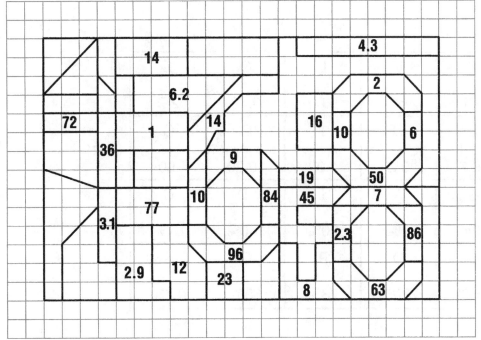

PERCENTS AND DECIMALS

ANIMAL DOCTOR

What do you call a sick crocodile?

Write each answer. Then find and cross out the answer in one of the boxes at the bottom. When you are finished, the remaining letters, written in order, will give you the answer to the question.

Write the percent as a decimal.

1. 50%   **2.** 102%   **3.** 9%   **4.** 27%

_____   _____   _____   _____

5. 0.4%   **6.** 18%   **7.** 20%   **8.** 4.1%

_____   _____   _____   _____

Write the decimal as a percent.

9. 0.02   **10.** 1.4   **11.** 1.04   **12.** 0.6

_____   _____   _____   _____

13. 0.007   **14.** 2   **15.** 0.81

_____   _____   _____

I	I	L	L	O	L
0.05	0.5	50	0.09	0.18	2.7
I	L	O	I	G	G
27.0	0.27	0.2	1.02	10.20	0.004
A	E	A	T	T	C
0.04	0.041	2%	20%	140%	81%
L	O	O	R	O	S
200%	104%	780%	78%	60%	0.7%

A sick crocodile is an ____ ____ ____ ____ ____

Name _____

FINDING CIRCUMFERENCE

THE LONG WAY AROUND

What is the distance around each figure? Round your answers to the nearest tenth.

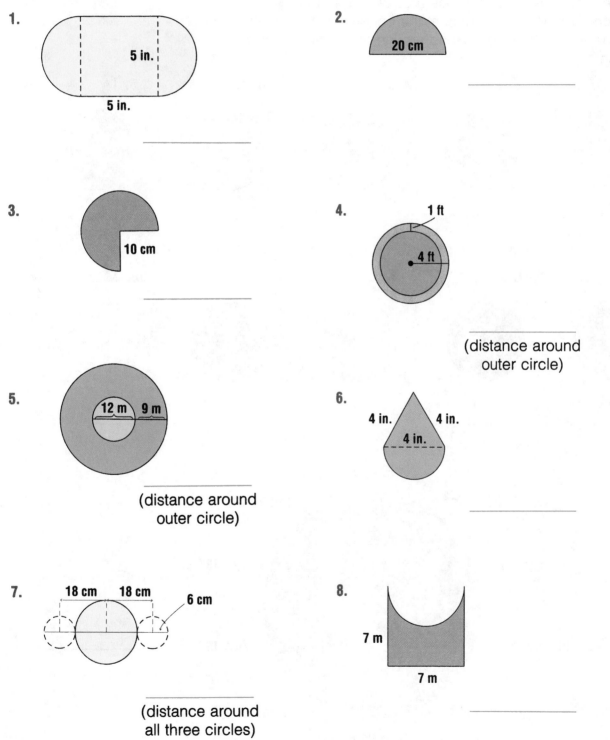

1. 5 in. / 5 in.

_____

2. 20 cm

_____

3. 10 cm

_____

4. 1 ft / 4 ft

_____
(distance around
outer circle)

5. 12 m / 9 m

_____
(distance around
outer circle)

6. 4 in. / 4 in. / 4 in.

_____

7. 18 cm / 18 cm / 6 cm

_____
(distance around
all three circles)

8. 7 m / 7 m

_____

AREA OF A CIRCLE

CIRCULAR REASONING

What kind of circle does a truck driver like?

Find the area of each circle. Use a calculator to help you. Then find each answer on the wheel at the bottom. Write the letter for each answer on the wheel. Read around the circle to answer the question.

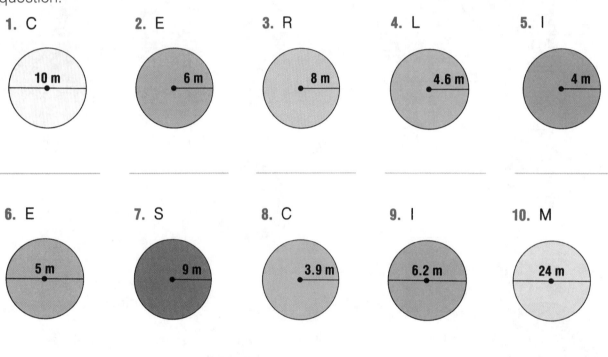

1. C — 10 m

2. E — 6 m

3. R — 8 m

4. L — 4.6 m

5. I — 4 m

6. E — 5 m

7. S — 9 m

8. C — 3.9 m

9. I — 6.2 m

10. M — 24 m

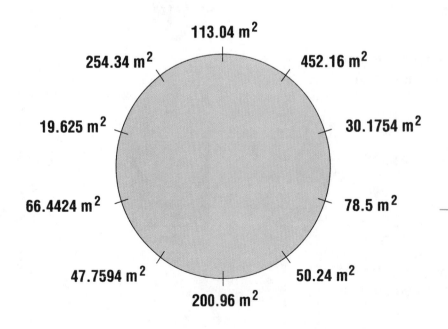

113.04 m²

254.34 m²   452.16 m²

19.625 m²   30.1754 m²

66.4424 m²   78.5 m²

47.7594 m²   50.24 m²

200.96 m²

DRAWING THREE-DIMENSIONAL FIGURES

KEEPING IT IN PERSPECTIVE

A. When you draw space figures, give a feeling of depth. Draw in **perspective.** As you look at a real object, like railroad tracks, for example, parallel lines *seem* to get closer in the distance. When you draw, instead of drawing lines parallel, make them get closer in the distance, too.

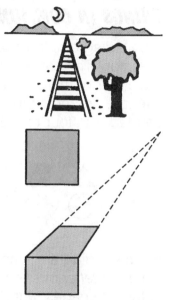

Here are some rules:

- Use parallel lines for any surface facing you, like the front of this prism.

- Make lines of surfaces that are receding (going away) from you slant towards each other slightly, like the top of this prism.

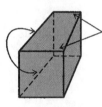

- Use the same slant for lines that go in the same direction, like these lines.

Complete these figures. Use perspective.

1. Rectangular prism

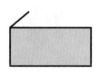

2. Triangular prism

3. Rectangular prism

B. Artists pick a point on their page and call it the vanishing point. All the receding parallel lines would meet at that point if they were extended far enough.

4. Draw a house in the space below. Pick a vanishing point. All receding parallel lines should meet at that point.

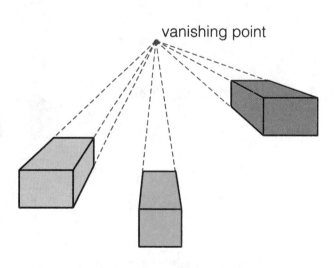

vanishing point

VOLUME OF A TRIANGULAR PRISM

THINGS IN COMMON

In this diagram, the shapes in the left circle all have something in common. They are all shaded. The shapes in the right circle all have four or more straight sides. In the middle, where the circles overlap, are the shapes which fit into both groups—they are shaded with four or more straight sides.

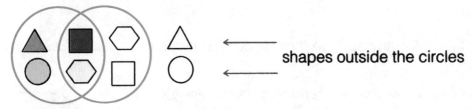

shapes outside the circles

For each diagram, decide where each shape should go. Then write the *letter* of each shape in the correct place. Write the letters of shapes that do not belong in any circle outside the circles.

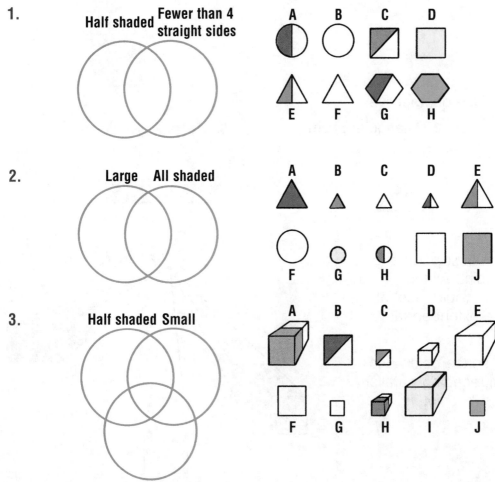

1.

Half shaded   Fewer than 4 straight sides

A   B   C   D

E   F   G   H

2.

Large   All shaded

A   B   C   D   E

F   G   H   I   J

3.

Half shaded   Small

A   B   C   D   E

F   G   H   I   J

Three-dimensional space figure

VOLUME OF A CYLINDER

CUTOUTS

Part of each of these three-dimensional figures has been cut out.
Find the volume of the remainder.

Assume that when you see a shape cut out of one face of a
figure, that cut goes through the entire figure. So, if you see a
circle cut out of one side, the cutout figure is a cylinder.

1.

3.

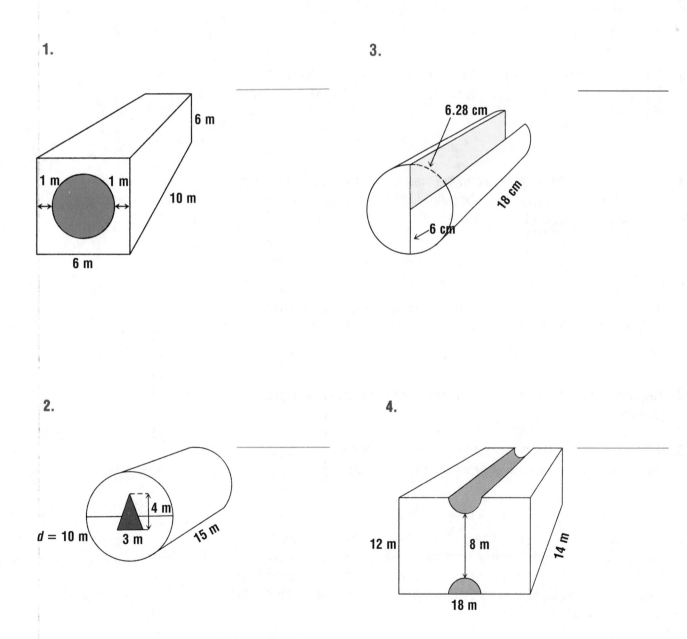

2.

4.

PROBABILITY

MIX AND MATCH

A. There are ten white socks and ten black socks in a drawer. Tony reaches into the drawer without looking.

1. What is the probability that he will pick a white sock?

_____

2. What is the probability that he will pick a black sock?

_____

3. Tony picks a white sock. Now he needs a match. What is the probability that he will pick a white sock from the drawer on the next try?

_____

4. Tony picks a black sock. Now he has one white sock and one black sock. What is the probability that he will have a pair of matching socks on the next try?

_____

5. Suppose the light is out so that Tony can't see the colors of the socks he picks. How many socks should he pick from the original 20 to be sure of getting at least one match?

_____

B. Assume that an equal number of people are born in each month.

6. If you try to guess the month someone is born in, what is the probability that you will guess the correct month?

_____

7. Suppose you are at a party with 35 people. You say that you can guarantee that you will find two people who were born in the same month if you can ask a certain number of people in what month he or she was born. What is the smallest number of people you can ask?

_____

LOGIC STATEMENTS

TRUE BLUE

The words *all, some,* or *no* (*none*) can be used to tell how two groups are related. Consider the following sentences:

> All sparrows are birds.
> Some birds are sparrows.
> Some birds are not sparrows.
> No birds are fish.

Notice that a sparrow is always a bird, but a bird is not always a sparrow (for example, a bird may be a robin). A bird is never a fish.

Write *true* or *false* for each statement.

1. All dogs are mammals. _Yes_

2. All mammals are dogs. _NO_

3. Some mammals are dogs. _Yes_

4. No tables are mammals. _NO_

5. All squares are quadrilaterals. _yes_

6. Some quadrilaterals are not squares. _yes_

7. Some squares are not polygons. _yes_

8. All numbers greater than 40 are numbers greater than 30. _yes_

9. Some numbers greater than 30 are numbers greater than 40. _yes_

10. No even numbers are odd numbers. _yes_

11. No odd numbers are numbers greater than 40. _NO_

12. Some even numbers are not numbers that end in 0, 2, 4, 6, or 8. _NO_

PROBLEM SOLVING

IT WORKS LIKE MAGIC

In a magic square, the sum of each row, column, and diagonal is the same. For example, here the sum is 150.

80	10	60
30	50	70
40	90	20

1. Look at the magic squares below. What do these squares and the one above have in common? (*Hint:* Think about simpler numbers.)

24	3	18
9	15	21
12	27	6

$2\frac{2}{3}$	$\frac{1}{3}$	$\frac{6}{3}$
1	$1\frac{2}{3}$	$2\frac{1}{3}$
$1\frac{1}{3}$	3	$\frac{2}{3}$

_____

_____

_____

_____

2. Now use the same pattern to make two more magic squares.

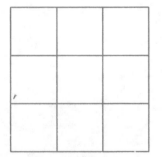

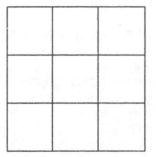

Name

GRAPHING ORDERED PAIRS

NAVAL DRILL

What kind of work does a dentist do in the navy?

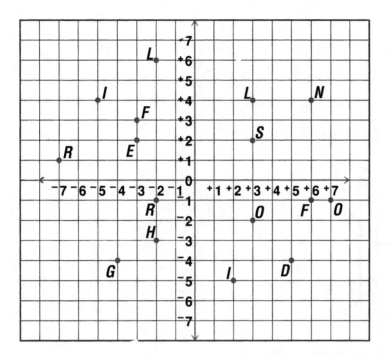

Find each of these points on the grid. Then write the letter for each point on the lines. You'll have the answer to the question.

($^+7$, $^-1$)   ($^-3$, $^+3$)   ($^+6$, $^-1$)   ($^+3$, $^+2$)

($^-2$, $^-3$)   ($^+3$, $^-2$)   ($^-7$, $^+1$)   ($^-3$, $^+2$)

($^+5$, $^-4$)   ($^-2$, $^-1$)   ($^-4$, $^+5$)   ($^-2$, $^+6$)

($^+3$, $^+4$)   ($^+2$, $^-5$)   ($^+6$, $^+4$)   ($^-4$, $^-4$)

GRAPHING TRANSFORMATIONS

LINE UP

1. Place each of these points on the grid below. Then connect the points.

   ($^{+}$2, $^{-}$4), ($^{+}$2, $^{-}$2), ($^{+}$2, $^{+}$1), ($^{+}$2, $^{+}$3)

2. What kind of figure do you get? _____

   In what direction does it go? _____

3. Compare the first numbers in each ordered pair. What do you

   notice about the numbers? _____

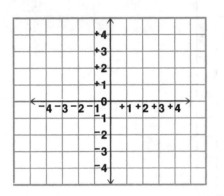

4. Name some other points that will make the same kind

   of line. _____

   Place these points on the grid.

5. What do you notice about the ordered pairs? _____

   ($^{-}$4, $^{+}$3), ($^{-}$2, $^{+}$3), ($^{+}$1, $^{+}$3), ($^{+}$3, $^{+}$3)

   What kind of figure do you get? _____

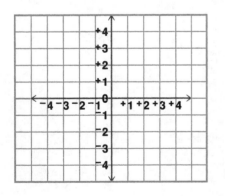

6. Name some other points that will make the same

   kind of line. _____

   Place these points on the grid.

ANSWER KEY

Name

WHOLE NUMBERS

WAY OUT IN SPACE

In 1772, an astronomer named Bode discovered something about the distances of planets from the Sun. The first seven planets in our solar system follow a number pattern.

1. Each number here is twice as big as the one before. Finish the pattern.

| 0 | 3 | 6 | 12 | 24 | 48 | 96 | 192 |

2. Next, add 4 to each number.

| 4 | 7 | 10 | 16 | 28 | 52 | 100 | 196 |
| Mercury | Venus | Earth | Mars | | Jupiter | Saturn | Uranus |

Look at the numbers that correspond to a planet. To find the distance of the planet from the sun, multiply the planet's number by 9,300,000 miles. Use your calculator and what you know about place value to find the product. Start with Earth, number 10.

10 × 9,300,000 = 93,000,000 miles from the Sun

Find the distances for the other planets.

3. Mercury	37,200,000 miles
4. Venus	65,100,000 miles
5. Earth	9,300,000 × 10 = 93,000,000 miles
6. Mars	148,800,000 miles
7. Jupiter	483,600,000 miles
8. Saturn	930,000,000 miles
9. Uranus	1,822,800,000 miles

What happened to the number between Mars and Jupiter? There is no planet there, but there is a group of asteroids. These rocky objects may be a planet that exploded. Also, Neptune and Pluto, the most distant planets, do not fit the pattern.

3

Name

DECIMALS

CROSS THAT BRIDGE

Choose the letter of the correct standard form for each item in the left column. The letters will spell the answer to this riddle.

What is the shortest bridge in the world?

B	1. six hundred and ten ten-thousandths	R 0.061
R	2. sixty-one thousandths	I 0.170
I	3. one hundred and seventy thousandths	F 0.2
D	4. fifty-three thousand and eleven hundred-thousandths	G 0.38
G	5. three hundred eighty thousandths	N 0.20360
E	6. two hundredths	Y 0.7710
O	7. one hundred and seventy-three thousandths	B 0.0610
F	8. two tenths	E 0.771
Y	9. seven thousand, seven hundred and ten ten-thousandths	G 0.380
D	10. thirty-eight hundredths	R 0.2036
U	11. seven hundredths	U 0.07
R	12. two thousand and thirty-six ten-thousandths	S 0.5301
N	13. twenty thousand, three hundred and sixty hundred-thousandths	D 0.53011
O	14. seven hundredths	O 0.173
S	15. five thousand, three hundred and one ten-thousandths	O 0.20
E	16. seven hundred and seventy-one thousandths	E 0.02

4

Name

MAKING BAR GRAPHS

AFTER-SCHOOL SPECIALTIES

Five students asked other students in their school, "What is your favorite after-school activity?" Here are the answers they got and the number of students who gave each answer.

Aaron's results: 3—chess club, 1—baseball, 1—music lessons, 1—dance lessons, 3—TV.

Bill's results: 2—swimming, 2—flute lessons, 2—TV, 2—computer club, 2—reading.

Sarah's results: 1—piano lessons, 2—acting club, 2—running, 2—gymnastics.

Kevin's results: 2—dance lessons, 1—acting club, 2—swimming, 2—baseball, 1—reading.

Jenny's results: 3—reading, 1—chess club, 2—computer club, 1—running, 1—baseball.

Make a graph to show this data.

1. Decide how you want to group the answers. Make a table showing your categories. **Answers will vary.**

2. Decide on the labels for the two axes and make your graph.

Possible answers:

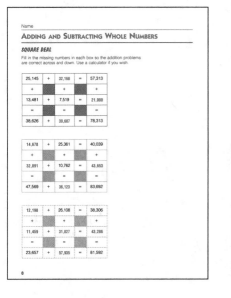

FAVORITE AFTER-SCHOOL ACTIVITIES

5

Name

ESTIMATING SUMS AND DIFFERENCES: ROUNDING

TARGET PRACTICE

Look at the target number in each circle. Then choose numbers from the box whose estimated sum or difference will be closest to the target. Write the addition or subtraction problem. For problems 1–5, choose two numbers.

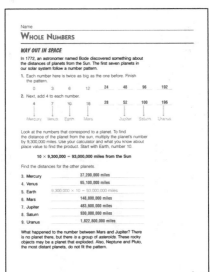

1. 700 — 537, 328, 89, 191 — 537 + 191
2. 5,000 — 956, 8,390, 2,738, 1,042 — 8,390 − 2,738
3. 9,000 — 1,379, 4,897, 6,101, 3,246 — 6,101 + 3,246
4. 13,000 — 2,871, 9,502, 12,430, 5,984 — 9,502 + 2,871
5. 20,000 — 11,769, 4,875, 2,958, 40,153, 32,066 — 32,066 − 11,769

For problems 6–7, you may choose more than two numbers and use more than one operation. **Answers may vary.**

6. 100,000 — 73,459, 5,652, 8,631, 47,268, 19,120 — 73,459 + 19,120 + 8,631
7. 0 — 25,898, 14,942, 3,758, 9,376, 40,645 — (25,898 + 14,942) − 40,645

6

Name

FRONT-END ESTIMATIONS: SUMS AND DIFFERENCES

PATTERN PICKS

How would you complete the following?

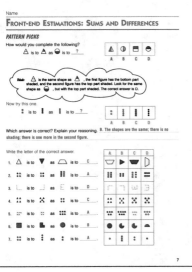

Now try this one.

Which answer is correct? Explain your reasoning. **B. The shapes are the same; there is no shading; there is one more in the second figure.**

Write the letter of the correct answer.

1. is to — as — is to — **C**
2. is to — as — is to — **A**
3. is to — as — is to — **D**
4. is to — as — is to — **C**
5. is to — as — is to — **A**
6. is to — as — is to — **B**
7. is to — as — is to — **A**

7

Name

ADDING AND SUBTRACTING WHOLE NUMBERS

SQUARE DEAL

Fill in the missing numbers in each box so the addition problems are correct across and down. Use a calculator if you wish.

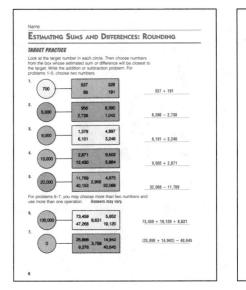

25,145	+	32,168	=	57,313
+		+		+
13,481	+	7,519	=	21,000
=		=		=
38,626	+	39,687	=	78,313

14,678	+	25,361	=	40,039
+		+		+
32,891	+	10,762	=	43,653
=		=		=
47,569	+	36,123	=	83,692

12,198	+	26,108	=	38,306
+		+		+
11,459	+	31,827	=	43,286
=		=		=
23,657	+	57,935	=	81,592

8

ANSWER KEY

Name
SOLVING EQUATIONS

EXTREME MEASURES

How do you measure poison ivy?

Solve each equation. Then draw a line to connect each pair of equations that have the same answer. (Use a ruler and draw a line between the dots.) The letters that are not crossed out will answer the riddle.

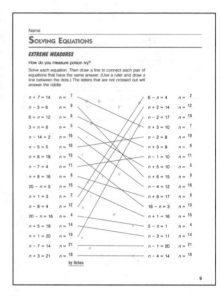

$n + 7 = 14$	$n =$ 7		$6 - n = 4$	$n =$	2
$n - 3 = 6$	$n =$ 9		$n + 2 = 14$	$n =$	12
$6 + n = 12$	$n =$ 6		$n - 2 = 17$	$n =$	19
$3 + n = 8$	$n =$ 5		$n - 3 = 10$	$n =$	7
$n - 14 = 2$	$n =$ 16		$n + 3 = 9$	$n =$	6
$n - 5 = 5$	$n =$ 10		$n - 2 = 8$	$n =$	10
$n + 6 = 19$	$n =$ 13		$n - 1 = 10$	$n =$	11
$n - 7 = 4$	$n =$ 11		$n - 5 = 10$	$n =$	5
$n + 8 = 16$	$n =$ 8		$n + 6 = 15$	$n =$	9
$20 - n = 5$	$n =$ 15		$n - 4 = 12$	$n =$	16
$n + 1 = 3$	$n =$ 2		$n - 4 = 4$	$n =$	8
$n - 8 = 4$	$n =$ 12		$16 - n = 3$	$n =$	13
$20 - n = 16$	$n =$ 4		$n + 1 = 16$	$n =$	15
$n + 5 = 19$	$n =$ 14		$5 - n = 1$	$n =$	4
$n + 1 = 20$	$n =$ 19		$n - 3 = 11$	$n =$	14
$n - 7 = 14$	$n =$ 21		$n - 1 = 20$	$n =$	21
$n + 3 = 21$	$n =$ 18		$n - 4 = 14$	$n =$	18

by itches

9

Name
MAKING LINE GRAPHS

COMMON QUALITIES

Here are two sets of numbers. Set A has odd numbers. Set B has even numbers:

A	B
1 3 45 9 11	2 66 4 82
25 7 87 99	6 10 18 48

Here the two sets overlap. The numbers that fit in set C where the sets overlap have something else in common. They are under 20. One set is even and under 20; one set is odd and under 20.

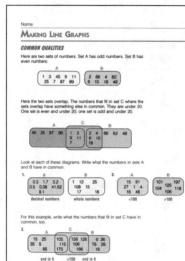

A: 45 25 87 99   C: 1 3 / 9 11 / 18   B: 2 66 82 48 / 6 10

Look at each of these diagrams. Write what the numbers in sets A and B have in common.

1. A: 0.3 1.7 3.2 / 0.5 0.06 41.62 / 9.1   C: 1 12 25 / 108 15 / 17 16
   decimal numbers   whole numbers

2. A: 15 91 / 27 1 4 / 16 46   B: 101 197 / 104 125 118 / 106
   <100   >100

For this example, write what the numbers that fit in set C have in common, too.

3. A: 15 25 / 35 5 / 65   C: 105 115 / 106 175 / 186   B: 156 126 / 106 16 / 6 36 / 26
   end in 5   >100   end in 6

10

Name
MENTAL MATH: MULTIPLYING

WHO'S WHO?

Anne, Beth, Charles, and David go to four different schools. The schools are Lincoln, Washington, Cleveland, and Roosevelt. Write the school that each student goes to. Use the clues below to help you.

Anne ____ Lincoln
Beth ____ Roosevelt
Charles ____ Cleveland
David ____ Washington

CLUES

1. The person who goes to Washington is a boy.
2. Charles used to go to Washington, but he changed schools this year.
3. Anne's school played softball against Roosevelt last month.
4. The student at Roosevelt is a girl.
5. Charles's cousin lives on the other side of town and goes to Lincoln.

You might find it helpful to fill in this chart as you get information about the students. Write X for no and ✓ for yes.

	Lincoln	Washington	Cleveland	Roosevelt
Anne	✓	X	X	X
Beth	X	X	X	✓
Charles	X	X	✓	X
David	X	✓	X	X

11

Name
MULTIPLYING

SCRAMBLED NUMBERS

Use the numbers in the box to write the factors and the product for each multiplication problem.

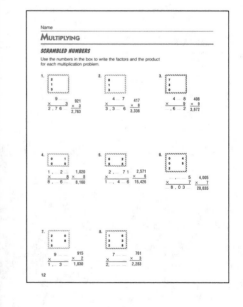

1. $\begin{array}{r} 9 \\ \times\ 3 \\ \hline 2,76 \end{array}$ $\begin{array}{r} 921 \\ \times\ 3 \\ \hline 2,763 \end{array}$

2. $\begin{array}{r} 4\ \ 7 \\ \times \\ \hline 3,\ 6 \end{array}$ $\begin{array}{r} 417 \\ \times\ 8 \\ \hline 3,336 \end{array}$

3. $\begin{array}{r} 4\ \ 8 \\ \times \\ \hline ,6\ \ 2 \end{array}$ $\begin{array}{r} 408 \\ \times\ 9 \\ \hline 3,672 \end{array}$

4. $\begin{array}{r} 1,\ \ 2 \\ \times\ \ 8 \\ \hline 8,\ \ 6 \end{array}$ $\begin{array}{r} 1,020 \\ \times\ 8 \\ \hline 8,160 \end{array}$

5. $\begin{array}{r} 2,\ \ 71 \\ \times \\ \hline 1,\ \ 4\ \ 6 \end{array}$ $\begin{array}{r} 2,571 \\ \times\ 6 \\ \hline 15,426 \end{array}$

6. $\begin{array}{r} ,\ \ 5 \\ \times\ \ 7 \\ \hline 8,03 \end{array}$ $\begin{array}{r} 4,005 \\ \times\ 7 \\ \hline 28,035 \end{array}$

7. $\begin{array}{r} 9 \\ \times \\ \hline 1,\ \ 3 \end{array}$ $\begin{array}{r} 915 \\ \times\ 2 \\ \hline 1,830 \end{array}$

8. $\begin{array}{r} 2 \\ \times \\ \hline 2, \end{array}$ $\begin{array}{r} 761 \\ \times\ 3 \\ \hline 2,283 \end{array}$

12

Name
MULTIPLYING GREATER NUMBERS

SERIOUS SERIES

These numbers follow a pattern. Each number is 12 times the number before it.

3   36   432   5,184

The next two numbers in the series are 62,208 (5,184 × 12) and 746,496 (62,208 × 12).

For each series, describe the pattern. Then fill in the next two numbers in each series. Use a calculator to help you. (Hint: Remember, not every series uses multiplication.)

1. 4 64 1,024 16,384   262,144   multiply by 16
2. 5 10 15 20 25   30   35   add 5
3. 3 51 867 14,739   250,563   multiply by 17
4. 23 184 1,472 11,776   94,208   753,664   multiply by 8
5. 3 8 40 45 225 230   1,150   1,155   add 5, multiply by 5
6. 6 13 91 98 686 693   4,851   4,858   add 7, multiply by 7
7. 12 10 660 658 43,428   43,426   2,866,116   subtract 2, multiply by 66
8. 19 10 90 81 729   720   6,480   subtract 9, multiply by 9
9. 124 248 496 992   1,984   3,968   multiply by 2
10. 375 75 600 300 2,400   2,100   16,800   subtract 300, multiply by 8

11. Write a series using a multiplication pattern. Give it to a friend to solve.

    Answers will vary.

12. Write a series with a pattern that uses addition or subtraction and multiplication. Give it to a friend to solve.

    Answers will vary.

13

Name
EXPONENTS

EXPONENTIALLY SPEAKING

Remember that you can use exponents to show multiplication when all the factors are the same.

$5 \times 5 \times 5 \times 5 = 5^4$

Now study the multiplication shown below.

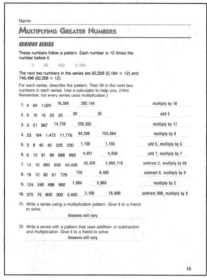

$(3 \times 3 \times 3) \times (3 \times 3) = 3 \times 3 \times 3 \times 3 \times 3$
$27 \times 9 = 243$

Write the exponents and the product in each problem. Then check that the products are correct. Use a calculator.

1. $\begin{array}{l} 3 \times 3 \times 3 = 3^3 \\ \times\ \ 3 \times 3 = \times\ 3^2 \\ \hline 3 \times 3 \times 3 \times 3 \times 3 = 3^5 \end{array}$ = $\begin{array}{r} 2\ 7 \\ \times\ 3\ 1 \\ \hline 2\ 4\ 3 \end{array}$

2. $\begin{array}{l} 4 \times 4 \times 4 = 4^3 \\ \times\ \ 4 \times 4 = \times\ 4^2 \\ \hline 4 \times 4 \times 4 \times 4 \times 4 = 4^5 \end{array}$ = $\begin{array}{r} 1\ 6 \\ \times\ 2\ 5\ 6 \\ \hline 4,0\ 9\ 6 \end{array}$

3. $\begin{array}{l} 2 \times 2 \times 2 = 2^3 \\ \times\ \ 2 \times 2 = \times\ 2^2 \\ \hline 2 \times 2 \times 2 \times 2 \times 2 = 2^5 \end{array}$ = $\begin{array}{r} 8 \\ \times\ 4 \\ \hline 6\ 4 \end{array}$

4. Can you write a rule that tells how to multiply two or more numbers in exponent form? What must be true about the base of each exponential number?

   The bases must be the same. To multiply, use the same base and add the exponents of the numbers.

Use your rule to write the answer to these problems. Then solve the factors and products in standard form. Check that the products are correct. Use a calculator.

5. $4^2 \times 4^2 = 4^4$ 16 × 16 = 256
6. $7^4 \times 7^3 = 7^7$ 2,401 × 343 = 823,543
7. $12^3 \times 12^2 = 12^5$ 1,728 × 144 = 248,832
8. $2^3 \times 2^3 \times 2^4 = 2^8$ 4 × 8 × 16 = 512

14

ANSWER KEY

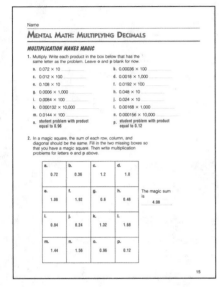

Mental Math: Multiplying Decimals

MULTIPLICATION MAKES MAGIC

1. Multiply. Write each product in the box below that has the same letter as the problem. Leave o and p blank for now.

a. 0.072 × 10
b. 0.00036 × 100
c. 0.012 × 100
d. 0.0018 × 1,000
e. 0.108 × 10
f. 0.0192 × 100
g. 0.0006 × 1,000
h. 0.048 × 10
i. 0.0084 × 100
j. 0.024 × 10
k. 0.000132 × 10,000
l. 0.00168 × 1,000
m. 0.0144 × 100
n. 0.000156 × 10,000
o. student problem with product equal to 0.96
p. student problem with product equal to 0.12

2. In a magic square, the sum of each row, column, and diagonal should be the same. Fill in the two missing boxes so that you have a magic square. Then write multiplication problems for letters o and p above.

a. 0.72	b. 0.36	c. 1.2	d. 1.8
e. 1.08	f. 1.92	g. 0.6	h. 0.48
i. 0.84	j. 0.24	k. 1.32	l. 1.68
m. 1.44	n. 1.56	o. 0.96	p. 0.12

The magic sum is 4.08

15

Multiplying Decimals

PUZZLING NUMBERS

Multiply. Then find each answer in the grid below. Write the word that matches each number in the grid.

1. is
```
   4.5
× 0.09
 0.405
```
2. funny
```
  11.36
× 9.98
113.3728
```
3. both
```
   0.89
× 4.39
 3.9071
```
4. have
```
   55.1
× 0.15
 8.265
```
5. A
```
   3.44
× 0.08
 0.2752
```
6. story
```
  0.578
×  7.7
 4.4506
```
7. pencil
```
  0.996
×  5.6
 5.5776
```
8. They
```
  0.566
× 0.45
 0.2547
```
9. a
```
  4.781
×  3.5
16.7335
```
10. like
```
 11.41
× 0.55
6.2755
```
11. a
```
  0.981
× 0.33
0.32373
```
12. point
```
  0.877
× 0.29
0.25433
```

0.2752 A	5.5776 pencil	0.405 is	6.2755 like
16.7335 a	113.3728 funny	4.4506 story.	0.2547 They
3.9071 both	8.265 have	0.32373 a	0.25433 point.

16

Relating Multiplication and Division

THE BLACK BOX

The black box contains a rule that changes numbers. Look at the numbers on the left. Then see what they become when they come out of the black box.

16 → 48
27 → 81
61 → 183

The rule inside the black box must be to multiply the number that goes in by 3.

Look at each of these examples. Write the rule. Then write the output for the last number. The rules may have more than one step, such as multiply by 3 and add 1.

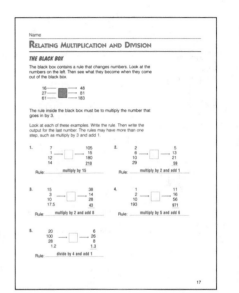

1.
```
7 → 105
1 → 15
12 → 180
14 → 210
```
Rule: multiply by 15

2.
```
6 → 5
10 → 21
29 → 59
```
Rule: multiply by 2 and add 1

3.
```
15 → 38
3 → 14
10 → 28
17.5 → 43
```
Rule: multiply by 2 and add 8

4.
```
1 → 11
2 → 16
10 → 56
193 → 971
```
Rule: multiply by 5 and add 6

5.
```
20 → 6
100 → 26
28 → 8
1.2 → 1.3
```
Rule: divide by 4 and add 1

17

Mental Math: Division Patterns

CODE BREAKER

The letters A to J stand for the numbers 0 to 9. Two letters stand for a two-digit number. Look at the problems below. They will help you break the code. Write the number that each letter stands for.

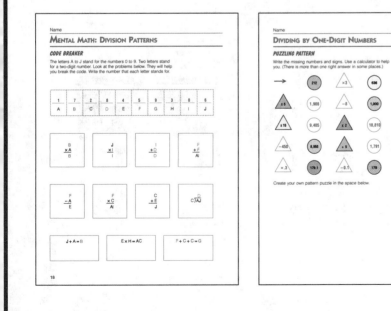

1	7	2	8	4	5	3	9	0	6
A	B	C	D	E	F	G	H	I	J

J + A = B

E × H = AC

F + C + C = G

18

Dividing by One-Digit Numbers

PUZZLING PATTERN

Write the missing numbers and signs. Use a calculator to help you. (There is more than one right answer in some places.)

Answers may vary.

→ 212   × 3 → 636   × 0.6 → 301.6

× 5 → 1,908   ÷ 6 → 1,900   × 0.33 → 527

× 15 → 9,405   ÷ 2 → 18,810   × 9.5 → 9,405

÷ 450 → 8,950   ÷ 5 → 1,791   ÷ 3 → 597

× .3 → 179.1   ÷ 0.1 → 179   + 33 → 212

Create your own pattern puzzle in the space below.

19

Dividing by Two-Digit Numbers

MISSING DIGITS

Which state is nicknamed the *Show Me State*? To find out, write the missing digits in each division problem. Then find the quotients below and write the correct letter in each space.

1. [S]
```
     1 2 R 1 4
3 2 ) 3 9 8
    3 2
    7 8
    6 4
    1 4
```

2. [O]
```
      1 5
5 7 ) 8 5 5
    5 7
    2 8 5
    2 8 5
```

3. [U]
```
      2 0 R 7
4 5 ) 9 0 7
    9 0
      0 7
      0
      7
```

4. [I]
```
      8 8 R 7
9 5 ) 8 3 6 7
    7 6 0
    7 6 7
    7 6 0
      7
```

5. [M]
```
      3 4 R 11
1 5 ) 5 8 9
    4 5
    1 3 9
    1 2 8
      1 1
```

6. [R]
```
      1 7 R 4
8 2 ) 1 3 9 8
    8 2
    5 7 8
    5 7 4
      4
```

The *Show Me State* is
M   I   S   S   O   U   R   I
34R11 88R7 12R14 12R14 15 20R7 17R4 80R7

20

ANSWER KEY

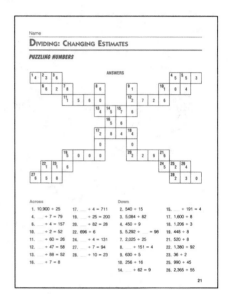

Name

DIVIDING: CHANGING ESTIMATES

PUZZLING NUMBERS

Across

1. 10,900 ÷ 25
4. ÷ 7 = 79
8. ÷ 4 = 157
10. ÷ 2 = 52
11. ÷ 60 = 26
12. ÷ 47 = 58
13. ÷ 88 = 52
16. ÷ 7 = 8

17. ÷ 4 = 711
19. ÷ 25 = 200
20. ÷ 82 = 28
22. 696 ÷ 6
24. ÷ 4 = 131
27. ÷ 7 = 94
28. ÷ 10 = 23

Down

2. 540 ÷ 15
3. 5,084 ÷ 82
4. 450 ÷ 9
5. 5,292 ÷ = 98
7. 2,025 ÷ 25
8. ÷ 151 = 4
9. 630 ÷ 5
10. 256 ÷ 16
14. ÷ 62 = 9

15. ÷ 191 = 4
17. 1,600 ÷ 8
18. 1,206 ÷ 3
19. 448 ÷ 8
20. 1,380 ÷ 92
23. 36 ÷ 2
25. 990 ÷ 45
26. 2,365 ÷ 55

21

Name

DIVIDING GREATER NUMBERS

SQUARE OFF

Use these numbers to fill in the square so that the division problems are correct across each row and down each column. One row is already filled in for you.

952 68 8,092 238 64,736 34

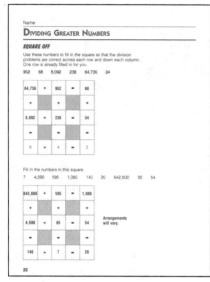

Fill in the numbers in this square.

7 4,590 595 1,080 140 20 642,600 95 54

Arrangements will vary.

22

Name

MULTIPLICATION AND DIVISION EQUATIONS

DEEP DEPTHS

Mt. Everest is the highest point on the earth. What is the lowest point? To find out, solve each equation. Write the letter above each question in the space below that matches *n*. (Write the letter everywhere there is a matching number.)

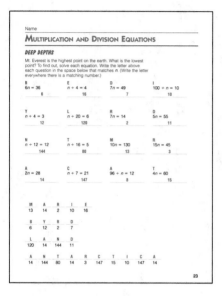

23

Name

MENTAL MATH: DIVIDING PATTERNS WITH POWERS OF 10

A SHRINKING VIOLET?

What gets shorter as it grows older?

Divide mentally. Then arrange the answers from least to greatest on the lines below. Write the letter for each problem above it. You'll have the answer to the riddle.

L 78.2 ÷ 100
E 14.33 ÷ 10
N 0.078 ÷ 10
N 122.3 ÷ 1,000
N 566.1 ÷ 1,000
U 2.03 ÷ 1,000
A 45.1 ÷ 100

R 3.2 ÷ 1,000
D 67.44 ÷ 100
B 0.113 ÷ 100
I 34.6 ÷ 1,000
C 1.8 ÷ 10
G 14.92 ÷ 100
A 0.3 ÷ 1,000

ANSWER:
A 0.0003
B 0.00113 U 0.00203 R 0.0032 N 0.0078 I 0.0346 N 0.1223 G 0.1492
C 0.18 A 0.451 N 0.5661 D 0.6744 L 0.782 E 1.433

24

Name

ZEROS IN THE QUOTIENT AND THE DIVIDEND

ZEROED OUT

Fill in the missing numbers in these division problems.

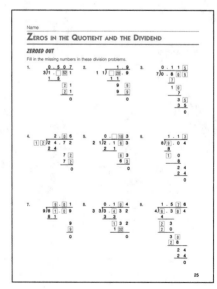

25

Name

DIVIDING DECIMALS BY DECIMALS

IF THE NUMBER FITS

For each problem, pick a divisor and quotient from the baskets to make a correct division problem. Use each number exactly once.

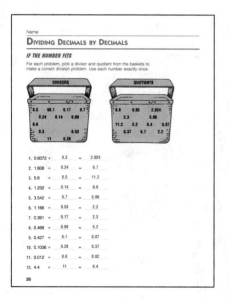

1. 0.6072 ÷ 0.3 = 2.024
2. 1.608 ÷ 0.24 = 6.7
3. 5.6 ÷ 0.5 = 11.2
4. 1.232 ÷ 0.14 = 8.8
5. 3.542 ÷ 0.7 = 5.06
6. 1.166 ÷ 0.53 = 2.2
7. 0.391 ÷ 0.17 = 2.3
8. 0.468 ÷ 0.09 = 5.2
9. 0.427 ÷ 6.1 = 0.07
10. 0.1036 ÷ 0.28 = 0.37
11. 0.012 ÷ 0.6 = 0.02
12. 4.4 ÷ 11 = 0.4

26

68

ANSWER KEY

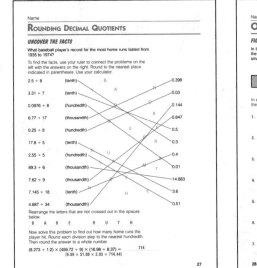

Name

ROUNDING DECIMAL QUOTIENTS

UNCOVER THE FACTS

What baseball player's record for the most home runs lasted from 1935 to 1974?

To find the facts, use your ruler to connect the problems on the left with the answers on the right. Round to the nearest place indicated in parentheses. Use your calculator.

2.5 ÷ 8	(tenth)	0.398
3.31 ÷ 7	(hundredth)	0.03
0.0976 ÷ 8	(hundredth)	0.144
6.77 ÷ 17	(thousandth)	0.847
0.25 ÷ 8	(hundredth)	0.5
17.8 ÷ 5	(tenth)	0.3
2.55 ÷ 5	(hundredth)	0.4
89.3 ÷ 6	(thousandth)	0.01
7.62 ÷ 9	(thousandth)	14.883
7.145 ÷ 18	(tenth)	3.6
4.887 ÷ 34	(thousandth)	0.51

Rearrange the letters that are not crossed out in the spaces below.

B A B E   R U T H

Now solve this problem to find out how many home runs the player hit. Round each division step to the nearest hundredth. Then round the answer to a whole number.

$(8.273 \div 1.2) \times (459.72 \div 9) \times (16.98 \div 8.37) =$ **714**
$[6.89 \times 51.08 \times 2.03 = 714.44]$

27

Name

ORDER OF OPERATIONS

FIGURE IT OUT

In the pattern below, there are two sets of figures. In each set, the figures are related in the same way: The second figure is a smaller version of the first.

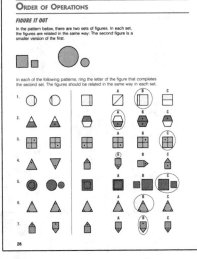

In each of the following patterns, ring the letter of the figure that completes the second set. The figures should be related in the same way in each set.

28

Name

MEAN, MEDIAN, MODE, AND RANGE

MAKE THE GRADE

Imagine that you are teacher for a day. Here are your students' scores on a math test. There are a possible 100 points on the test. What grade will you give each student?

In order to be fair, you should use some statistics to help you. You can calculate the mean, mode, median, and range of the scores. Make up a scale telling what scores get a passing grade (A, B, C, or D) and what scores fail (F).

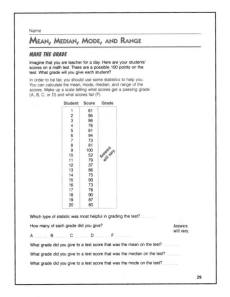

Student	Score	Grade
1	81	
2	95	
3	86	
4	76	
5	81	
6	94	
7	73	
8	81	
9	100	
10	52	Answers
11	79	will vary.
12	37	
13	86	
14	75	
15	93	
16	73	
17	76	
18	90	
19	87	
20	80	

Which type of statistic was most helpful in grading the test? _____

How many of each grade did you give? _____   Answers will vary.

A _____ B _____ C _____ D _____ F _____

What grade did you give to a test score that was the mean on the test? _____

What grade did you give to a test score that was the median on the test? _____

What grade did you give to a test score that was the mode on the test? _____

29

Name

FACTORS AND GREATEST COMMON FACTOR

DAFFY DINO DEFINITION

Which dinosaur was the fastest?

Using a ruler, draw a line from the numbers on the left to their greatest common factor (GCF) on the right. The letters that remain will answer the question. Write the letters in order, reading from top to bottom, in the blank below.

1.	12, 28	P	15
2.	40, 4, 2	R	4
3.	75, 25, 15		12
4.	12, 24	O	20
5.	30, 40	N	16
6.	45, 60, 75	T	2
7.	60, 80		5
8.	16, 48, 80	S	10
9.	12, 18	U	8
10.	14, 21	R	6
11.	45, 72, 90		9
12.	64, 24	U S	7

The fastest dinosaur was the   **PRONTOSAURUS**

30

Name

MULTIPLES AND LEAST COMMON MULTIPLE

CAREFUL ARRANGEMENTS

How can you arrange four 3s so the value is 36? Here is one way:

$(3 + 3) \times (3 + 3) \to 6 \times 6 = 36$

In the puzzles below, you can use any operation signs you choose, as well as decimal points or fractions. **Answers will vary.**

1. Arrange four 5s so the value is 100.
   $(5 + 5) \times (5 + 5)$

2. Arrange four 9s so the value is 153.
   $(9 + 9) \times 9 - 9$

3. Arrange four 9s so the value is 20.
   $99 \div 9 + 9$

4. Arrange four 5s so the value is 2.5.
   $(.5 \times .5) \times (5 + 5)$ or $(5 \times 5) \div (5 + 5)$

5. Arrange four 3s so the value is 34.
   $33 + \frac{3}{3}$

6. Arrange five 3s so the value is 17.
   $33 + \frac{33}{3} \div 3$

7. Arrange five 2s so the value is 0.
   $(22 + 2) \times (2 - 2)$

31

Name

FRACTIONS AND EQUIVALENT FRACTIONS

FRACTURED FAIRY TALE

What fairy tale is about some talking vegetables?

Write an equivalent fraction for each of the following fractions. Then find the answers under the lines below. Write the letter for each fraction on the line. The first letter has been filled in.

32

69

ANSWER KEY

MIXED NUMBERS

CURIOUS CODE

In each problem, the numbers have been replaced by letters. Each letter always stands for the same number. It may be a 1-digit number or a 2-digit number. All answers are fractions or mixed numbers written in simplest terms.

Use the clue and logic to decide what number each letter stands for. Write the 12 letters and numbers on the lines below.

CLUE: L = 2

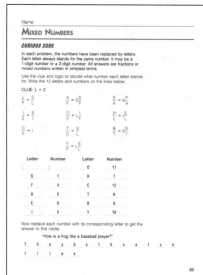

Letter	Number	Letter	Number
L	2	O	11
S	1	H	7
F	4	C	12
A	5	T	8
E	9	B	6
I	3	Y	10

Now replace each number with its corresponding letter to get the answer to this riddle:

"How is a frog like a baseball player?"

T h e y   b o t h   c a t c h

f l i e s

33

COMPARING AND ORDERING

DAFFYNITIONS

To answer each riddle, write each group of fractions and mixed numbers in order from least to greatest. Write the letter of each number below the line.

What do you call the place where school lunches are made?

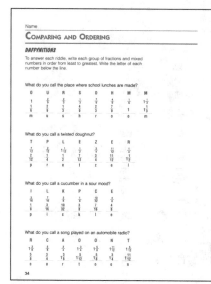

O   U   R   S   O   H   M   M

m   u   s   h   r   o   o   m

What do you call a twisted doughnut?

T   P   L   E   Z   E   R

p   r   e   t   z   e   l

What do you call a cucumber in a sour mood?

I   L   K   P   C   E

p   i   c   k   l   e

What do you call a song played on an automobile radio?

R   C   A   O   O   N   T

c   a   r   t   o   o   n

34

ROUNDING FRACTIONS AND MIXED NUMBERS

FRACTION ACTION

You can use mental math to decide if a fraction is greater than $\frac{1}{2}$.

Example: Is $\frac{3}{4} > \frac{1}{2}$?

Double the numerator. $2 \times 3 = 6$

If the result is greater than the denominator, $6 > 5$

the fraction is greater than $\frac{1}{2}$.   so $\frac{3}{4} > \frac{1}{2}$

Is the fraction greater than $\frac{1}{2}$? Write YES or NO.

1. $\frac{3}{5}$ YES   2. $\frac{2}{5}$ NO   3. $\frac{4}{11}$ NO   4. $\frac{7}{10}$ NO

5. $\frac{5}{10}$ YES   6. $\frac{4}{10}$ NO   7. $\frac{18}{27}$ YES   8. $\frac{11}{17}$ YES

In the maze below, find a path from START to END. Use only fractions that are greater than $\frac{1}{2}$. You may go up, down, across, or diagonally, but do not cross a number more than once.

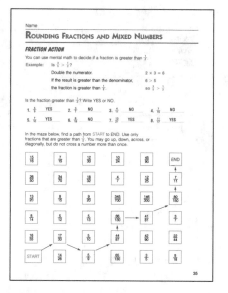

35

CUSTOMARY UNITS OF LENGTH

GOING TO GREAT LENGTH

Arrange these items from longest to shortest. (You'll have to find a way to compare their lengths first, since they are all given in different units.)

Remember, a mile is 5,280 feet.

White Sea—Baltic Canal, Soviet Union	141 mi
Golden Gate Bridge, California	1,400 yd
Simplon I Tunnel, Switzerland to Italy	12.3 mi
Sydney Harbor Bridge, Australia	19,800 in.
Panama Canal, Panama	50.7 mi
Mackinac Straits Bridge, Michigan	45,600 in.
Verrazano Narrows Bridge, New York	4,260 ft
Welland Canal, Canada	49,280 yd
Suez Canal, Egypt	100.6 mi
New River Gorge Bridge, West Virginia	1,700 ft
Rokko Tunnel, Japan	633,600 in.
Northern Line Subway Tunnel, England	30,488 yd

1.	White Sea—Baltic Canal, Soviet Union	141 mi
2.	Suez Canal, Egypt	100.6 mi
3.	Panama Canal, Panama	50.7 mi
4.	Welland Canal, Canada	49,280 yd
5.	Northern Line Subway Tunnel, England	30,488 yd
6.	Simplon I Tunnel, Switzerland to Italy	12.3 mi
7.	Rokko Tunnel, Japan	633,600 in.
8.	Verrazano Narrows Bridge, New York	4,260 ft
9.	Golden Gate Bridge, California	1,400 yd
10.	Mackinac Straits Bridge, Michigan	45,600 in.
11.	New River Gorge Bridge, West Virginia	1,700 ft
12.	Sydney Harbor Bridge, Australia	19,800 in.

36

FRACTIONS, MIXED NUMBERS, AND DECIMALS

CROSSHATCH

Try to find a decimal in columns 3 and 4 that matches each fraction or mixed number in columns 1 and 2. When you find a match, cross out both the numbers and the letters next to them. A match may be found in any row. An example is done for you.

COLUMN 1	COLUMN 2	COLUMN 3	COLUMN 4
C $1\frac{1}{10}$	A	U 0.34	A
E $\frac{73}{1000}$	P $4\frac{89}{100}$	E 6.25	T 0.11
O	E	B	D 1.1
T $\frac{7}{8}$	I $7\frac{99}{100}$	E	A 4.089
B	R	Q 9.855	E
T $\frac{5}{10}$	I	N	E 9.12
N $\frac{17}{50}$	N $9\frac{1}{2}$	A 7.099	S 7.5
E	O $\frac{7}{1000}$	M	T
H $3\frac{5}{100}$	T $\frac{7}{10}$	L 1.02	L 2.8

Now write the letters that are left in each column on the line. Unscramble each group of letters to spell a familiar mathematical word.

COLUMN 1	ETTNH	TENTH
COLUMN 2	PINOT	POINT
COLUMN 3	UEGAL	EQUAL
COLUMN 4	TAESL	LEAST

37

ADDING FRACTIONS: UNLIKE DENOMINATORS

THE MISSING LINK

Fill in the boxes with the missing numbers. Each fraction in these problems should be in its simplest form.

1. $\frac{1}{\boxed{}}$ $+ \frac{1}{4}$ = $\frac{3}{4}$

2. $\frac{1}{6}$ $+ \frac{1}{3}$ = $\frac{1}{2}$

3. $\frac{3}{\boxed{}}$ $+ \frac{1}{4}$ = $\frac{5}{8}$

4. $\frac{5}{16}$ $+ \frac{3}{16}$ = $\frac{1}{2}$

5. $\frac{1}{3}$ $+ \frac{1}{4}$ = $\frac{7}{12}$

6. $\frac{1}{3}$ $+ \frac{5}{6}$ = $1\frac{1}{6}$

7. $\frac{3}{10}$ $+ \frac{9}{10}$ = $1\frac{1}{5}$

8. $\frac{3}{4}$ $+ \frac{5}{8}$ = $1\frac{3}{8}$

38

70

ANSWER KEY

Subtracting Fractions: Unlike Denominators

Name

PICK A PAIR

1. Which two numbers have a sum of exactly $\frac{1}{2}$?

$$\frac{1}{3} \quad \frac{1}{4} \quad \frac{1}{6} \quad \frac{1}{5} \quad \frac{1}{8}$$

$$\frac{1}{4} \quad \frac{1}{4}$$

2. Which two numbers have a difference of $\frac{1}{4}$?

$$\frac{2}{3} \quad \frac{1}{3} \quad \frac{3}{8} \quad \frac{5}{16} \quad \frac{3}{10}$$

$$\frac{3}{8} \quad \frac{1}{4}$$

3. Which two numbers have the sum that is closest to 1?

$$\frac{9}{10} \quad \frac{1}{3} \quad \frac{1}{20} \quad \frac{9}{8} \quad \frac{1}{8}$$

$$\frac{9}{10} \quad \frac{1}{8}$$

4. Which two numbers have the difference that is closest to $\frac{1}{2}$?

$$\frac{7}{10} \quad \frac{3}{4} \quad \frac{9}{16} \quad \frac{1}{12} \quad$$

$$\frac{9}{16} \quad \frac{1}{12}$$

39

Adding Mixed Numbers

Name

WHAT'S FOR DINNER?

What's the best kind of pie to take on a picnic?

Write the answer to each problem. Then find each answer at the bottom. Write the letter for each problem on the line to answer the riddle.

P $3\frac{1}{4}$ + $2\frac{3}{8}$ = $5\frac{5}{8}$

O $4\frac{1}{4}$ + $3\frac{3}{12}$ = $7\frac{1}{4}$

H $12\frac{1}{8}$ + $3\frac{1}{2}$ = $15\frac{5}{8}$

F $1\frac{3}{4}$ + $1\frac{1}{8}$ = $2\frac{5}{8}$

Y $4\frac{2}{8}$ + $1\frac{3}{8}$ = $5\frac{5}{8}$

S $9\frac{1}{4}$ + $1\frac{3}{8}$ = $10\frac{5}{8}$

I $1\frac{11}{12}$ + $4\frac{1}{2}$ = $6\frac{5}{12}$

E $1\frac{4}{8}$ + $1\frac{1}{8}$ = $2\frac{7}{24}$

L $3\frac{1}{4} + 2\frac{1}{2} + 6\frac{1}{4}$ = 12

O $7\frac{1}{10} + 1\frac{1}{4} + 2\frac{1}{2}$ = $11\frac{1}{5}$

$10\frac{1}{4}$	$15\frac{5}{8}$	$11\frac{3}{8}$	$7\frac{1}{4}$
S	H	O	O

$2\frac{5}{8}$	12	$5\frac{5}{8}$
F	L	Y

$5\frac{5}{8}$	$6\frac{5}{12}$	$2\frac{7}{24}$
P	I	E

40

Subtracting Mixed Numbers

Name

PART MAGIC

Fill in the missing fractions in each magic square. The sum should be the same across, down, and on the diagonals of each square.

1.

$\frac{1}{12}$		$\frac{1}{4}$
$\frac{1}{3}$	$\frac{1}{6}$	0
$\frac{1}{12}$	$\frac{1}{6}$	$\frac{1}{4}$

Magic Sum $\frac{1}{2}$

2.

$\frac{3}{4}$	$\frac{1}{2}$	$\frac{1}{4}$
0	$\frac{1}{2}$	1
$\frac{3}{4}$	$\frac{1}{2}$	$\frac{1}{4}$

Magic Sum $1\frac{1}{2}$

3.

$\frac{4}{5}$	$1\frac{4}{5}$	
$\frac{3}{5}$	1	$1\frac{2}{5}$
$1\frac{3}{5}$	$\frac{1}{5}$	$1\frac{1}{5}$

Magic Sum 3

41

Problem Solving

Name

YOU BE THE TEACHER

Imagine that you are a teacher and that the problems below were solved by some of your students. Look at each problem. If the answer is correct, place a ✔ next to it. If the answer is incorrect, give the correct solution. Then explain the error so you can help your students to do better work. The first one is done for you.

Error explanations may vary. Check that child's descriptions are appropriate.

1. Student A

$$856$$
$$723$$
$$+ 624$$
$$\overline{2,193} \quad 2,203$$

Error: Did not regroup correctly

2. Student B

145.56 + 236.3 + 99.732 = 1,166.47

481.592

Error: Did not align decimal points.

3. Student C

A box of paper plates costs $1.79 and a box of cups costs $0.79. How much change do you receive from $20 if you buy six boxes of plates and fives boxes of cups?

Error: $5.31 ✔

None.

4. Student D

A dozen pencils costs $2.79 and a dozen pens costs $5.95. How much more do 3 dozen pens cost than 2 dozen pencils?

Error: $23.43 $12.27

used the wrong operation

5. Student E

A box can hold 15 books. How many boxes will you need to pack 110 books?

Error: 7 boxes 8 boxes

did not take into account the remaining 5 books after filling 7 boxes

6. Student F

6 × (15 + 4) − 2 = 92

112

Error: did not do operation in parentheses first

42

Multiplying Factors

Name

FLIPPER AND FRIENDS

What kind of animal is a dolphin or porpoise?

Match the multiplication problem on the left to the answer on the right. Write the letter of the answer in the box. The letters will spell the answer to the question.

S 1. ☐ $\frac{1}{4} \times \frac{2}{5}$ W 1

M 2. ☐ $\frac{2}{3} \times \frac{1}{3}$ A $\frac{7}{10}$

A 3. ☐ $\frac{7}{8} \times \frac{4}{5}$ A 4

L 4. ☐ 3 × $\frac{2}{3}$ S $\frac{1}{6}$

L 5. ☐ $\frac{3}{4} \times \frac{1}{4}$ H $\frac{7}{8}$

W 6. ☐ 4 × $\frac{1}{4}$ L $\frac{1}{10}$

H 7. ☐ $\frac{7}{16} \times \frac{2}{5}$ E $\frac{21}{40}$

A 8. ☐ 6 × $\frac{2}{3}$ L 2

L 9. ☐ $\frac{2}{3} \times \frac{3}{4}$ M $\frac{1}{9}$

E 10. ☐ $\frac{7}{10} \times \frac{3}{4}$ L $\frac{3}{5}$

43

Multiplying Mixed Numbers

Name

POSITIVELY PYRAMIDS

What numbers do cooks like?

Multiply the number at the top of each triangle by the numbers in the circles in the second row. Write the answers in simplest form in the bottom circles. Then arrange the answers in order from least to greatest. Copy the letter from each answer.

Numbers in order: $\frac{9}{32}$ $\frac{19}{64}$ $\frac{14}{45}$ $\frac{9}{20}$ $1\frac{7}{32}$ $1\frac{7}{8}$ $2\frac{7}{27}$ $2\frac{2}{5}$ $3\frac{1}{36}$ $5\frac{41}{64}$

Letters: M I X E D N U M B E R S

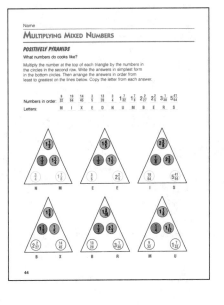

44

ANSWER KEY

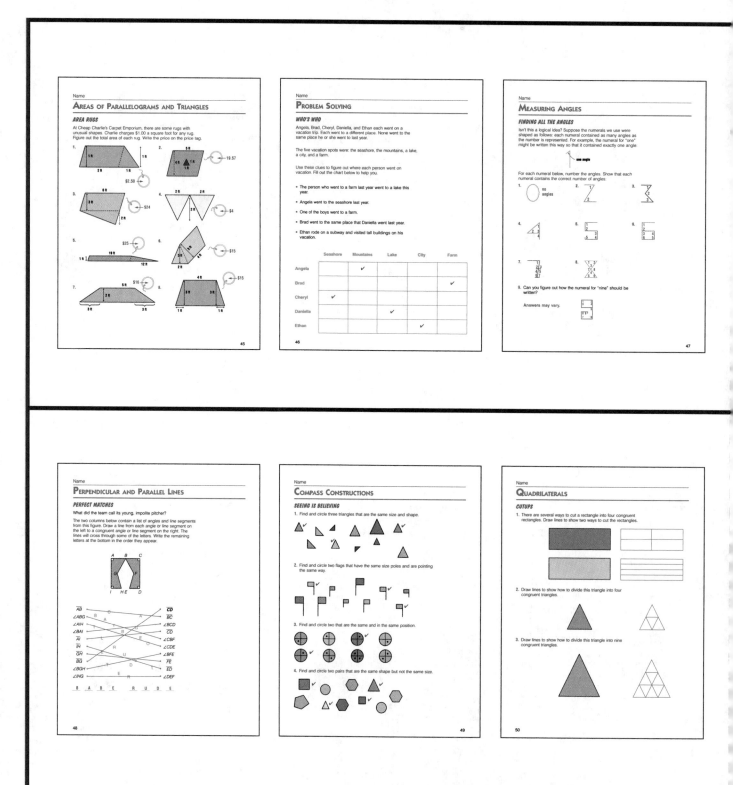

Name
AREAS OF PARALLELOGRAMS AND TRIANGLES
AREA RUGS

At Cheap Charlie's Carpet Emporium, there are some rugs with unusual shapes. Charlie charges $1.00 a square foot for any rug. Figure out the total area of each rug. Write the price on the price tag.

1. $2.50
2. $19.57
3. $24
4. $4
5. $25
6. $15
7. $16
8. $15

Name
PROBLEM SOLVING
WHO'S WHO

Angela, Brad, Cheryl, Daniella, and Ethan each went on a vacation trip. Each went to a different place. None went to the same place he or she went to last year.

The five vacation spots were: the seashore, the mountains, a lake, a city, and a farm.

Use these clues to figure out where each person went on vacation. Fill out the chart below to help you.

• The person who went to a farm last year went to a lake this year.

• Angela went to the seashore last year.

• One of the boys went to a farm.

• Brad went to the same place that Daniella went last year.

• Ethan rode on a subway and visited tall buildings on his vacation.

	Seashore	Mountains	Lake	City	Farm
Angela		✓			
Brad					✓
Cheryl	✓				
Daniella			✓		
Ethan				✓	

Name
MEASURING ANGLES
FINDING ALL THE ANGLES

Isn't this a logical idea? Suppose the numerals we use were shaped as follows: each numeral contained as many angles as the number is represented. For example, the numeral for "one" might be written this way so that it contained exactly one angle.

For each numeral below, number the angles. Show that each numeral contains the correct number of angles:

1. no angles
2.
3.
4.
5.
6.
7.
8.

9. Can you figure out how the numeral for "nine" should be written?

Answers may vary.

Name
PERPENDICULAR AND PARALLEL LINES
PERFECT MATCHES

What did the team call its young, impolite pitcher?

The two columns below contain a list of angles and line segments from this figure. Draw a line from each angle or line segment on the left to a congruent angle or line segment on the right. The lines will cross through some of the letters. Write the remaining letters at the bottom in the order they appear.

$\overline{AB}$ — $\overline{CD}$
$\angle ABG$ — $\overline{BC}$
$\angle AIH$ — $\angle BCD$
$\angle BAI$ — $\overline{CD}$
$\overline{AI}$ — $\angle CBF$
$\overline{IH}$ — $\angle CDE$
$\overline{GH}$ — $\angle BFE$
$\overline{BG}$ — $\overline{FE}$
$\angle BGH$ — $\overline{ED}$
$\angle IHG$ — $\angle DEF$

B A B E   R U D E

Name
COMPASS CONSTRUCTIONS
SEEING IS BELIEVING

1. Find and circle three triangles that are the same size and shape.

2. Find and circle two flags that have the same size poles and are pointing the same way.

3. Find and circle two that are the same and in the same position.

4. Find and circle two pairs that are the same shape but not the same size.

Name
QUADRILATERALS
CUTUPS

1. There are several ways to cut a rectangle into four congruent rectangles. Draw lines to show two ways to cut the rectangles.

2. Draw lines to show how to divide this triangle into four congruent triangles.

3. Draw lines to show how to divide this triangle into nine congruent triangles.

ANSWER KEY

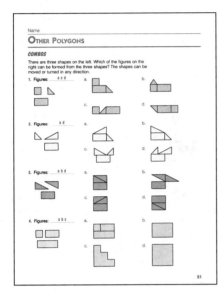

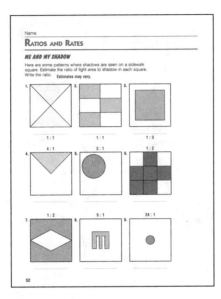

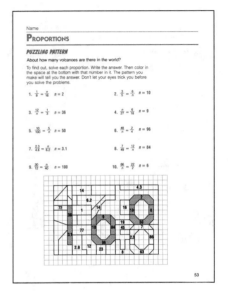

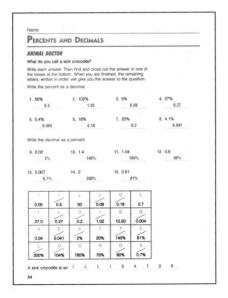

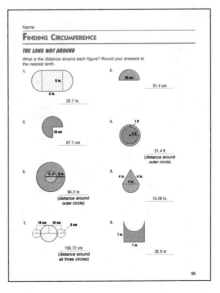

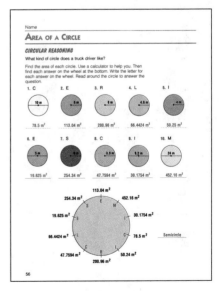

ANSWER KEY

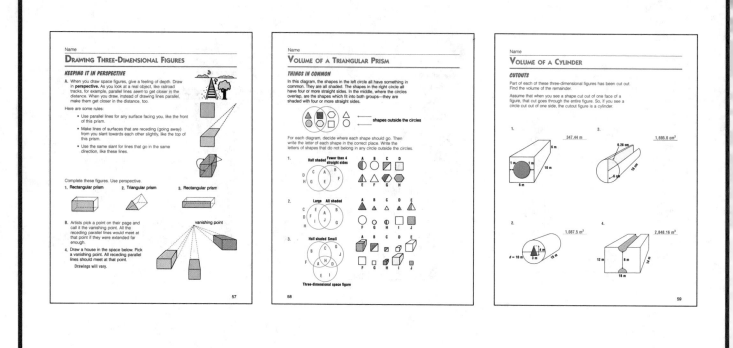

Drawing Three-Dimensional Figures

KEEPING IT IN PERSPECTIVE

A. When you draw space figures, give a feeling of depth. Draw in **perspective.** As you look at a real object, like railroad tracks, for example, parallel lines seem to get closer in the distance. When you draw, instead of drawing lines parallel, make them get closer in the distance, too.

Here are some rules:
- Use parallel lines for any surface facing you, like the front of this prism.
- Make lines of surfaces that are receding (going away) from you slant towards each other slightly, like the top of this prism.
- Use the same slant for lines that go in the same direction, like these lines.

Complete these figures. Use perspective.

1. Rectangular prism 2. Triangular prism 3. Rectangular prism

B. Artists pick a point on their page and call it the vanishing point. All the receding parallel lines would meet at that point if they were extended far enough.

vanishing point

4. Draw a house in the space below. Pick a vanishing point. All receding parallel lines should meet at that point.

Drawings will vary.

57

Volume of a Triangular Prism

THINGS IN COMMON

In this diagram, the shapes in the left circle all have something in common. They are all shaded. The shapes in the right circle all have four or more straight sides. In the middle, where the circles overlap, are the shapes which fit into both groups—they are shaded with four or more straight sides.

shapes outside the circles

For each diagram, decide where each shape should go. Then write the letter of each shape in the correct place. Write the letters of shapes that do not belong in any circle outside the circles.

1. Half shaded / Fewer than 4 straight sides
2. Large / All shaded
3. Half shaded / Small

Three-dimensional space figure

58

Volume of a Cylinder

CUTOUTS

Part of each of these three-dimensional figures has been cut out. Find the volume of the remainder.

Assume that when you see a shape cut out of one face of a figure, that cut goes through the entire figure. So, if you see a circle cut out of one side, the cutout figure is a cylinder.

1. 347.44 m
3. 1,695.6 cm³
2. 1,087.5 m³
4. 2,848.16 m³

59

Probability

MIX AND MATCH

A. There are ten white socks and ten black socks in a drawer. Tony reaches into the drawer without looking.

1. What is the probability that he will pick a white sock? $\frac{1}{2}$

2. What is the probability that he will pick a black sock? $\frac{1}{2}$

3. Tony picks a white sock. Now he needs a match. What is the probability that he will pick a white sock from the drawer on the next try? $\frac{9}{19}$

4. Tony picks a black sock. Now he has one white sock and one black sock. What is the probability that he will have a pair of matching socks on the next try? 1 or 100%

5. Suppose the light is out so that Tony can't see the colors of the socks he picks. How many socks should he pick from the original 20 to be sure of getting at least one match? 11

B. Assume that an equal number of people are born in each month.

6. If you try to guess the month someone is born in, what is the probability that you will guess the correct month? $\frac{1}{12}$

7. Suppose you are at a party with 35 people. You say that you can guarantee that you will find two people who were born in the same month if you can ask a certain number of people in what month he or she was born. What is the smallest number of people you can ask? 13

60

Logic Statements

TRUE BLUE

The words all, some, or no (none) can be used to tell how two groups are related. Consider the following sentences:

All sparrows are birds.
Some birds are sparrows.
Some birds are not sparrows.
No birds are fish.

Notice that a sparrow is always a bird, but a bird is not always a sparrow (for example, a bird may be a robin). A bird is never a fish.

Write true or false for each statement.

1. All dogs are mammals. T
2. All mammals are dogs. F
3. Some mammals are dogs. T
4. No tables are mammals. T
5. All squares are quadrilaterals. T
6. Some quadrilaterals are not squares. T
7. Some squares are not polygons. F
8. All numbers greater than 40 are numbers greater than 30. T
9. Some numbers greater than 30 are numbers greater than 40. T
10. No even numbers are odd numbers. T
11. No odd numbers are numbers greater than 40. F
12. Some even numbers are not numbers that end in 0, 2, 4, 6, or 8. F

61

Problem Solving

IT WORKS LIKE MAGIC

In a magic square, the sum of each row, column, and diagonal is the same. For example, here the sum is 150.

80	10	60
30	50	70
40	90	20

1. Look at the magic squares below. What do these squares and the one above have in common? (Hint: Think about simpler numbers.)

24	3	18
9	15	21
12	27	6

$2\frac{3}{4}$	$\frac{1}{4}$	$2\frac{1}{4}$
1	$1\frac{3}{4}$	
$1\frac{1}{3}$	3	

All are multiples of a square with these numbers:

8	1	6
3	5	7
4	9	2

2. Now use the same pattern to make two more magic squares.

Answers will vary but should follow the pattern.

62

ANSWER KEY

Graphing Ordered Pairs

Name _____

NAVAL DRILL

What kind of work does a dentist do in the navy?

Find each of these points on the grid. Then write the letter for each point on the lines. You'll have the answer to the question.

O	F	F	S
($^+$7, $^-$1)	($^-$3, $^+$3)	($^+$6, $^-$1)	($^+$3, $^+$2)

H	D	R	E
($^-$2, $^-$3)	($^+$3, $^-$2)	($^-$7, $^+$1)	($^-$3, $^+$2)

D	R	I	L
($^+$5, $^-$4)	($^-$2, $^-$1)	($^-$4, $^+$5)	($^-$2, $^+$6)

L	I	N	G
($^+$3, $^+$4)	($^+$2, $^-$5)	($^+$6, $^+$4)	($^-$4, $^-$4)

63

Graphing Transformations

Name _____

LINE UP

1. Place each of these points on the grid below. Then connect the points.
   ($^+$2, $^-$4), ($^+$2, $^-$2), ($^+$2, $^+$1), ($^+$2, $^+$3)
2. What kind of figure do you get? __a straight line__
   In what direction does it go? __vertical__
3. Compare the first numbers in each ordered pair. What do you notice about the numbers? __same number__

4. Name some other points that will make the same kind of line. __Answers will vary, any pairs where the first number is the same__
   Place these points on the grid.
5. What do you notice about the ordered pairs? __Second number is the same in each pair.__
   ($^-$4, $^+$3), ($^-$2, $^+$3), ($^+$1, $^+$3), ($^+$3, $^+$3)
   What kind of figure do you get? __Straight line__

6. Name some other points that will make the same kind of line. __any pairs where the second number is the same__
   Place these points on the grid.

64

McGraw-Hill Consumer Products

The skills taught in school are now available at home!
These award-winning software titles meet school guidelines and are based on
The McGraw-Hill Companies classroom software titles.

MATH GRADES 1 & 2

These math programs are a great way to teach and reinforce skills used in everyday situations. Fun, friendly characters need help with their math skills. Everyone's friend, Nubby the stubby pencil, will help kids master the math in the Numbers Quiz show. Foggy McHammer, a carpenter, needs some help building his playhouse so that all the boards will fit together! Julio Bambino's kitchen antics will surely burn his pastries if you don't help him set the clock timer correctly! We can't forget Turbo Tomato, a fruit with a passion for adventure, who needs help calculating his daredevil stunts.

Math Grades 1 & 2 use a tested, proven approach to reinforcing your child's math skills while keeping him or her intrigued with Nubby and his collection of crazy friends.

> TITLE
> Grade 1: Nubby's Quiz Show
> Grade 2: Foggy McHammer's Treehouse

MISSION MASTERS™ MATH AND LANGUAGE ARTS

The Mission Masters™—Pauline, Rakeem, Mia, and T.J.—need your help. The Mission Masters™ are a team of young agents working for the Intelliforce Agency, a high-level cooperative whose goal is to maintain order on our rather unruly planet. From within the agency's top secret Command Control Center, the agency's central computer, M5, has detected a threat...and guess what—you're the agent assigned to the mission!

MISSION MASTERS™ MATH GRADES 3, 4, & 5

This series of exciting activities encourages young mathematicians to challenge themselves and their math skills to overcome the perils of villains and other planetary threats. Skills reinforced include: analyzing and solving real-world problems, estimation, measurements, geometry, whole numbers, fractions, graphs, and patterns.

> TITLE
> Grade 3: Mission Masters™ Defeat Dirty D!
> Grade 4: Mission Masters™ Alien Encounter
> Grade 5: Mission Masters™ Meet Mudflat Moe

MISSION MASTERS™ LANGUAGE ARTS GRADES 3, 4, & 5

This series invites children to apply their language skills to defeat unscrupulous characters and to overcome other earthly dangers. Skills reinforced include: language mechanics and usage, punctuation, spelling, vocabulary, reading comprehension, and creative writing.

> TITLE
> Grade 3: Mission Masters™ Freezing Frenzy
> Grade 4: Mission Masters™ Network Nightmare
> Grade 5: Mission Masters™ Mummy Mysteries

BASIC SKILLS BUILDER K to 2 – THE MAGIC APPLEHOUSE

At the Magic Applehouse, children discover that Abigail Appleseed runs a deliciously successful business selling apple pies, tarts, and other apple treats. Enthusiasm grows as children join in the fun of helping Abigail run her business. Along the way they'll develop computer and entrepreneurial skills to last a lifetime. They will run their own business – all while they're having bushels of fun!

TITLE
Basic Skills Builder –The Magic Applehouse

TEST PREP – SCORING HIGH

This grade-based testing software will help prepare your child for standardized achievement tests given by his or her school. Scoring High specifically targets the skills required for success on the Stanford Achievement Test (SAT) for grades three through eight. Lessons and test questions follow the same format and cover the same content areas as questions appearing on the actual SAT tests. The practice tests are modeled after the SAT test-taking experience with similar directions, number of questions per section, and bubble-sheet answer choices.

Scoring High is a child's first-class ticket to a winning score on standardized achievement tests!

TITLE
Grades 3 to 5: Scoring High Test Prep
Grades 6 to 8: Scoring High Test Prep

SCIENCE

Mastering the principles of both physical and life science has never been so FUN for kids grades six and above as it is while they are exploring McGraw-Hill's edutainment software!

TITLE
Grades 6 & up: Life Science
Grades 8 & up: Physical Science

REFERENCE

The National Museum of Women in the Arts has teamed with McGraw-Hill Consumer Products to bring you this superb collection available for your enjoyment on CD-ROM.

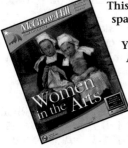

This special collection is a visual diary of 200 women artists from the Renaissance to the present, spanning 500 years of creativity.

You will discover the art of women who excelled in all the great art movements of history. Artists who pushed the boundaries of abstract, genre, landscape, narrative, portrait, and still-life styles; as well as artists forced to push the societal limits placed on women through the ages.

TITLE
Women in the Arts

Most titles for Windows 3.1™, Windows '95™ & '98™, and Macintosh™.

Visit us on the Internet at:

www.MHkids.com

Or call 800-298-4119 for your local retailer.

McGraw-Hill Consumer Products

All our workbooks meet school curriculum guidelines and correspond to
The McGraw-Hill Companies classroom textbooks.

SPECTRUM SERIES

DOLCH Sight Word Activities

The DOLCH Sight Word Activities Workbooks use the classic Dolch list of 220 basic vocabulary words that make up from 50% to 75% of all reading matter that children ordinarily encounter. Since these words are ordinarily recognized on sight, they are called *sight words*. Volume 1 includes 110 sight words. Volume 2 covers the remainder of the list. Over 160 pages.

TITLE	ISBN	PRICE
Grades K-1 Vol. 1	1-57768-429-X	$9.95
Grades K-1 Vol. 2	1-57768-439-7	$9.95

GEOGRAPHY

Full-color, three-part lessons strengthen geography knowledge and map reading skills. Focusing on five geographic themes including location, place, human/environmental interaction, movement, and regions. Over 150 pages. Glossary of geographical terms and answer key included.

TITLE	ISBN	PRICE
Gr 3, Communities	1-57768-153-3	$7.95
Gr 4, Regions	1-57768-154-1	$7.95
Gr 5, USA	1-57768-155-X	$7.95
Gr 6, World	1-57768-156-8	$7.95

MATH

Features easy-to-follow instructions that give students a clear path to success. This series has comprehensive coverage of the basic skills, helping children to master math fundamentals. Over 150 pages. Answer key included.

TITLE	ISBN	PRICE
Grade 1	1-57768-111-8	$6.95
Grade 2	1-57768-112-6	$6.95
Grade 3	1-57768-113-4	$6.95
Grade 4	1-57768-114-2	$6.95
Grade 5	1-57768-115-0	$6.95
Grade 6	1-57768-116-9	$6.95
Grade 7	1-57768-117-7	$6.95
Grade 8	1-57768-118-5	$6.95

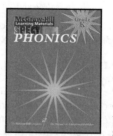

PHONICS

Provides everything children need to build multiple skills in language. Focusing on phonics, structural analysis, and dictionary skills, this series also offers creative ideas for using phonics and word study skills in other language arts. Over 200 pages. Answer key included.

TITLE	ISBN	PRICE
Grade K	1-57768-120-7	$6.95
Grade 1	1-57768-121-5	$6.95
Grade 2	1-57768-122-3	$6.95
Grade 3	1-57768-123-1	$6.95
Grade 4	1-57768-124-X	$6.95
Grade 5	1-57768-125-8	$6.95
Grade 6	1-57768-126-6	$6.95

SPECTRUM SERIES – continued

READING

This full-color series creates an enjoyable reading environment, even for below-average readers. Each book contains captivating content, colorful characters, and compelling illustrations, so children are eager to find out what happens next. Over 150 pages. Answer key included.

TITLE	ISBN	PRICE
Grade K	1-57768-130-4	$6.95
Grade 1	1-57768-131-2	$6.95
Grade 2	1-57768-132-0	$6.95
Grade 3	1-57768-133-9	$6.95
Grade 4	1-57768-134-7	$6.95
Grade 5	1-57768-135-5	$6.95
Grade 6	1-57768-136-3	$6.95

SPELLING

This full-color series links spelling to reading and writing and increases skills in words and meanings, consonant and vowel spellings, and proofreading practice. Over 200 pages. Speller dictionary and answer key included.

TITLE	ISBN	PRICE
Grade 1	1-57768-161-4	$7.95
Grade 2	1-57768-162-2	$7.95
Grade 3	1-57768-163-0	$7.95
Grade 4	1-57768-164-9	$7.95
Grade 5	1-57768-165-7	$7.95
Grade 6	1-57768-166-5	$7.95

WRITING

Lessons focus on creative and expository writing using clearly stated objectives and pre-writing exercises. Eight essential reading skills are applied. Activities include main idea, sequence, comparison, detail, fact and opinion, cause and effect, and making a point. Over 130 pages. Answer key included.

TITLE	ISBN	PRICE
Grade 1	1-57768-141-X	$6.95
Grade 2	1-57768-142-8	$6.95
Grade 3	1-57768-143-6	$6.95
Grade 4	1-57768-144-4	$6.95
Grade 5	1-57768-145-2	$6.95
Grade 6	1-57768-146-0	$6.95
Grade 7	1-57768-147-9	$6.95
Grade 8	1-57768-148-7	$6.95

TEST PREP
From the Nation's #1 Testing Company

Prepares children to do their best on current editions of the five major standardized tests. Activities reinforce test-taking skills through examples, tips, practice, and timed exercises. Subjects include reading, math, and language. Over 150 pages. Answer key included.

TITLE	ISBN	PRICE
Grade 1	1-57768-101-0	$8.95
Grade 2	1-57768-102-9	$8.95
Grade 3	1-57768-103-7	$8.95
Grade 4	1-57768-104-5	$8.95
Grade 5	1-57768-105-3	$8.95
Grade 6	1-57768-106-1	$8.95
Grade 7	1-57768-107-X	$8.95
Grade 8	1-57768-108-8	$8.95

Visit us on the Internet at:
www.MHkids.com

Or call 800-298-4119 for your local retailer.

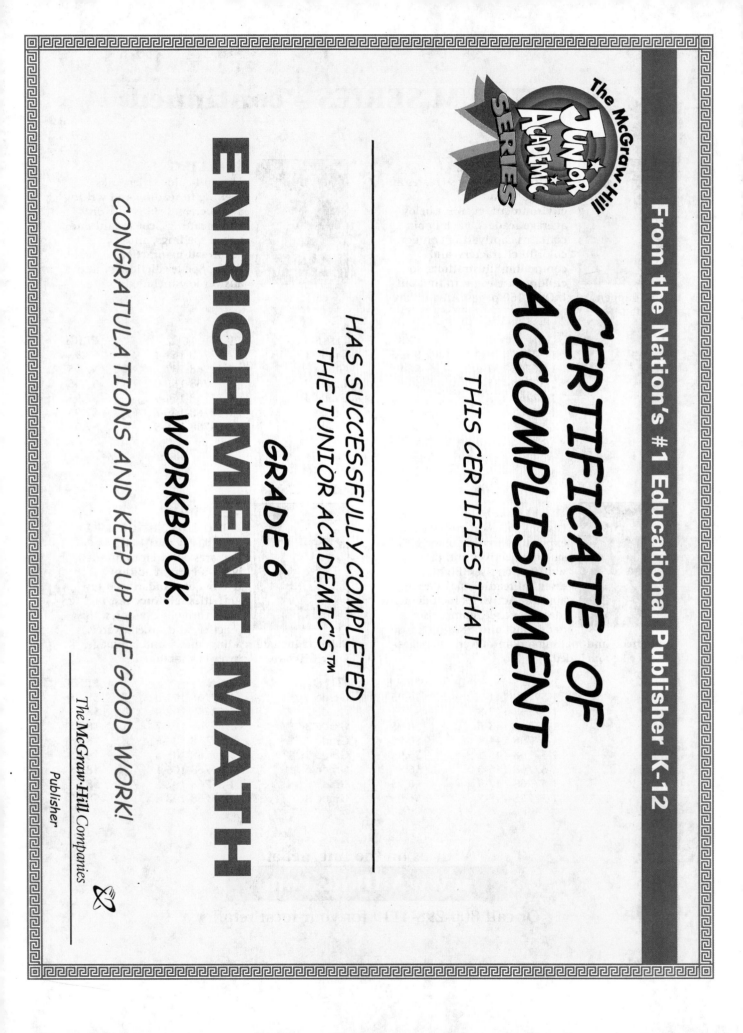